Number Six: Environmental History Series

MARTIN V. MELOSI, *General Editor*

MARSHES OF THE OCEAN SHORE

MARSHES
OF THE
OCEAN SHORE
Development of an
Ecological Ethic

by

JOSEPH V. SIRY

Texas A&M University Press

COLLEGE STATION

Copyright © 1984 by Joseph V. Siry

Library of Congress Cataloging in Publication Data

Siry, Joseph V. (Joseph Vincent), 1949–
    Marshes of the ocean shore.

    (Environmental history series ; no. 6
    Bibliography: p.
    Includes index.
    1. Tidemarsh ecology.   2. Coastal zone management.
I. Title.   II. Series.
QH541.5.S24S57   1983   333.91′8′0973   82-45899
ISBN 0-89096-150-6

Manufactured in the United States of America
FIRST EDITION

To my family,
especially my mother and father

Contents

List of Illustrations

MARSHES OF THE OCEAN SHORE

1

Introduction

All the rivers run into the sea,
Yet the sea is not full.
From whence the rivers come
Thither they return again.
—Ecclesiastes 1:7

A natural reciprocal nurturing of ocean and earth creates abundant wildlife in coastal wetlands. Here, salt and fresh water flow together in tidal marshes, creating rare shorelines of unsurpassed natural fertility by converting solar energy into food. Swelling tidal fluctuations recycle vital nutrients that encourage rapid vegetational growth followed by quick decay. The estuarine life cycle sustains fertile meadows and ponds for countless bacteria, insects, fish, birds, and mammals. Both native and migratory wildlife thrive on the tide-mulched marsh-grass fields. An early observer remarked: "All organic life is beautifully and variedly adjusted to the conditions of its environment, but it is doubtful if in any other zone of the organic world the accommodations are more exquisitely ordered than in the marshes of the ocean shore."[1]

Coastal bodies of water where streams of rivers flow into the ocean are called estuaries.[2] Protected from the outer seas' surge by barrier islands, sand dunes, submerged reefs, peninsulas, or rocky promontories, estuaries are bordered by vast acres of low-lying, water-tolerant vegetation broadly classified as coastal wetlands. Salt marshes, tidelands, swamps, or sloughs are transitional areas of wetland vegetation that flourish between

[1] James Morris, "The Estuary: One of Nature's Keystones," in *Essays in Social Biology*, ed. Bruce Wallace, I, 243–45; Nathaniel Southgate Shaler, *Sea and Land*, p. 250.
[2] W. M. Cameron and Donald Pritchard define an estaury as "a semi-enclosed body of water having free connection to the open sea, within which sea water is measurably diluted with fresh water deriving from land drainage" ("Estuaries," in *The Sea*, ed. M. N. Hill [New York: John Wiley and Sons, 1963], p. 306).

the deep waters of streams, creeks, or estuaries and the dry lands of the coastal plain.

Shoreline wetlands are distinguished according to the causes of their submergence. Rivers, through their seasonal or annual cycles of rain and flood, create swamps and overflowed lands. In coastal regions these swamps and overflowed lands frequently merge with tidelands inundated by the predictable ebbs and floods of tides in estuaries, bays, and oceans.[3] Tidelands, lying along the shore between the usual reaches of the tides, include both herbaceous tidal marshes and related but unvegetated mud flats. According to ecologists, these tidal flats and marshes sustain enormous numbers of wildlife.[4]

An integral part of estuaries, tidal marshes are those portions of coastal wetlands formed by tidal action and sedimentation in certain river mouths and bays. Throughout history, these marshlands have been assigned varied significances, exemplified by a number of biblical accounts. In the book of Genesis, salt and fresh marshes are necessary elements in the creation of the earth, a theme reflecting Babylonian and Egyptian creation stories. The book of Exodus depicts marshes as a wasteland through which the captive Israelites must pass before entering the wilderness beyond. The marshes are both a refuge for the fleeing slaves and a quagmire for Pharaoh's pursuing army. Marshes and rivers are described in the book of Ezekiel as divine agents of cultural restoration. Since biblical times, the transformation of wetlands lying astride principal river mouths has been part of the ancient western dream of conquest over nature's physical forces.[5]

While long viewed as wastelands, these encroaching tidal marshes along estuarine creeks are now regarded by ecologists as the earth's most productive natural areas. The evolution and adaptation of marine life to fresh-water conditions occurred in the coastal wetlands that crowd many rivers' shores. Estuarine wetlands are also major centers of human civiliza-

---

[3]Francois D. Uzes, *Chaining the Land: A History of Surveying in California*, pp. 131–33. There are 88,633 statute miles of shores that are exposed to tidal variation, as compared with 12,383 miles of total seacoast. Rachel Carson ("Our Ever Changing Shore," *Holiday Magazine*, July, 1958) uses a figure of 53,677 miles of tidally affected coast.

[4]Eugene P. Odum, "The Role of Tidal Marshes in Estuarine Production," *Conservationist*, June–July, 1961, p. 13.

[5]Raymond Brown et al., eds. *The Jerome Biblical Commentary*, II, 346, 348, 362; John R. Stilgoe, "Jack-o'-Lanterns to Surveyors: The Secularization of Landscape Boundaries," *Environmental Review* 1 (January, 1976): 14–31.

tions. The challenge to continuing wetland conservation in the United States lies in balancing marsh preservation with urban expansion. Balancing is difficult because the coasts are the nation's most densely populated regions, while the estuarine tidal marshes and wetlands are the earth's most productive ecosystems. The change in national temper from viewing coastal wetlands as wastelands to appreciating them as preserves of marine biological abundance significantly deepened and improved scientific understanding of life's physical evolution and endurance.

This study focuses on tidal marshes as parts of the estuarine system in order to explore how a more accurate and thorough scientific comprehension of matter altered conventional beliefs, legal interpretations, and government policies concerning the protection of seashores. Developments in biochemistry and ecology have contributed new evidence on the tidal wetlands' role in estuarine and coastal productivity. Recent ecological evidence has challenged the assumption that tidal marshes and mud flats are wastelands. Rather, river mouths, the nexus of interior valleys and coastal plains, are natural capillaries in the earth's circulatory system, wherein fresh water returns to the sea. Furthermore, on an immediate pragmatic level, commercial fisheries depend on the numerous plants and animals thriving along coastal wetlands.[6] Marshland preservation advocates, today, partially base their arguments on recent research depicting tidal marshes and adjacent estuaries as some of the earth's most fertile natural communities.

All varieties of marshland provide food for tiny organisms, but the tidal marsh–estuarine ecosystem supports greater numbers because the waters draining the tidal marshes flow into oceans. Shrimp, oysters, and crabs are three marine fisheries that flourish in estuaries. The abundant source of food generated by the tidal marshes also attracts adult flounder, sea bass, and juvenile migratory fish, especially salmon, herring, and shad. Estuarine wetlands sustain commercial fisheries, numerous birds, and various wildlife.[7]

One reason that estuarine marshes are more fecund than interior wetlands is the constant ebb and flood of the tides, which distribute food to all

[6] Eugene P. Odum, *Fundamentals of Ecology*, pp. 352–62; Robert M. Ingle, "The Life of an Estuary," *Scientific American*, May, 1956, pp. 64–68; John Teal and Mildred Teal, *Life and Death of the Salt Marsh*, pp. 39–52, 122–261; Peggy Weyburn, *The Edge of Life*.

[7] Institute of Ecology, *Man in the Living Environment: A Report on Global Problems*, pp. 244–60.

parts of the river mouth while pumping wastes out to sea. The movement of the tides and the flow of the rivers cause water to move in opposite directions within estuaries. Besides carrying salt water to the interior, the circulation patterns of estuaries hold the river's valuable nutrients. Where fresh water is diluted by salt water, nutrients and minerals are trapped and recycled by estuarine plant and animal life.[8] In addition to sustaining wildlife, tidal marshes and their surrounding estuaries have a limited capacity to filter and degrade complex natural waste products. The amount of food available to fish or wildlife and the ability of estuaries to recycle nutrient-rich wastes are largely related to the extent of marshes and mud flats within the river mouth.

Overall, historic patterns in American development of estuaries reveal conflicting assumptions concerning the use of natural resources, changing national attitudes about government-sponsored scientific research, and persistent prejudices against the least understood sections of the coast, the tidal marshes. Legally, tidelands comprise a public trust of state-owned lands for the promotion of fishing and hunting and the protection of navigation. As publicly owned commons, tidelands provide nursery grounds for fisheries, recreational access to the shore, and waste disposal sites for civic and industrial pollution.

Because of the accessibility of most estuaries to inland regions and oceans, trade and commerce have always been a principal use of quiescent portions of river mouths. Frequently siltation has threatened estuarine navigability with encroaching tidal marshes. Depending on the particular locale, tidal marshes are apt to convert the mud flats along rivers, estuaries, or coastal seas into wetland meadows because grasses or sedges trap additional silt and can survive extended periods of submergence. Despite the recognizably crucial role that tidal marshes play in the natural productivity of estuaries and coastal seas, their hindrance of commerce encourages the

---

[8]Odum, "Role of Tidal Marshes," pp. 13–14; Wesley Marx, *The Frail Ocean*, pp. 71–109, 148–49, 170–79; D. S. Ranwell, *Ecology of Salt Marshes and Sand Dunes*, pp. 201–13; G. Tyler Miller, Jr., *Living in the Environment: Concepts, Problems and Alternatives*, pp. 173–78, 269; R. S. K. Barnes and J. Green, eds., *The Estuarine Environment*, pp. 1–9. Where the nation's river valleys meet the sea, considerable amounts of pollution-laden sediment are deposited, and within those deposits historical layering of pollution reflects each different watershed's land-use patterns over time (See Edward Goldberg, "Pollution History of Estuarine Sediments," *Oceanus* 19 [Fall, 1976]: 18–26).

reclamation of coastal wetlands.[9] Because this reclamation threatens estuarine productivity, the continued enhancement of estuarine resources remains the most difficult problem for coastal conservation.

Discussions concerning the use of estuarine resources involve the distinctions between conservation and preservation. Although both words suggest the protection of natural resources, their precise meanings and the history of their associated political movements in America differ. Preservation of historic sites and picturesque landscapes developed in nineteenth-century cities through private contributions and local civic initiative. Such preservation of natural or cultural resources implies a qualitative judgment of the aesthetic merits and historical meaning associated with landscapes, structures, objects, or artifacts in order to maintain a sense of tradition.[10] The social commitment to a policy of environmental preservation presupposes that a culture partially finds its identity in relation to the functional integrity of the environment.

Conservation of natural or cultural resources, implies, on the other hand, a quantitative judgment of the future needs and productive potential of economically vital raw materials and manufactured goods. The word *conservation* was derived from conservators, who were East Indian foresters appointed by the British during the nineteenth century. Two advisers to Theodore Roosevelt, Gifford Pinchot and W. J. McGee, borrowed the word to apply to their Progressive policies for the use of natural resources in a manner that would bring maximum benefits to the largest number of citizens.[11]

It is tempting to equate the twin impulses of preservation and conservation with a dualism inherent in the environmental tradition of western

[9]L. Eugene Cronin, "The Role of Man in Estuarine Processes," in *Man's Impact on the Environment*, ed. Thomas Detwyler, pp. 266–94; U.S. Congress, Senate, *National Estuarine Pollution Study, Report of the Secretary of the Interior*. Senate Document 91–58, 91st Cong., 2d sess., March 25, 1970, pp. 1–263; John H. Davis, "Influence of Man upon Coast Lines," in *Man's Role in Changing the Face of the Earth*, ed. William L. Thomas, Jr., pp. 504–21.

[10]Roderick Nash, ed., *The American Environment*, pp. 37–38, 68–78; Donald Worster, ed., *American Environmentalism: The Formative Period, 1860–1915*, pp. 1–184; Roderick Nash, "The American Conservation Movement," in *Forums in History*, pp. 9–10, 507; Charles H. Hosmer, *The Presence of the Past: A History of the Preservation Movement*, pp. 41–122.

[11]Gifford Pinchot, *The Fight for Conservation*, pp. 4, 43–44, 48–49; Samuel P. Hays, *Conservation and the Gospel of Efficiency: The Progressive Conservation Movement, 1890–1920*, pp. 1–26, 91–121, 261–79; Nash, *American Environment*, pp. 37–52.

civilization, characterized by historical geographer Clarence Glacken as an ambivalent regard for nature as both useful and beautiful.[12] Equating preservation with beauty and conservation with use oversimplifies the subtle patterns resulting from the economic development of estuarine resources. Fisheries in tidal marshes, for example, may be conserved directly by regulating fishing, but, dependent as they are on the biotic fecundity of the wetlands, they may also benefit indirectly from policies protecting the beauty of marshes by making parks.

On the other hand, conservation for use and preservation for aesthetic reasons may meet head-on in conflicts over the proper form of development. The beaches of the outer coasts, long valued for their aesthetic and recreational appeal, depend upon the deposition of sand in river mouths. However, these very sediments, responsible for the vast stretches of dunes and beaches characterizing the Atlantic, Gulf, and southern California Pacific coasts, may impede navigation and thus be detrimental to policies designed to conserve waterways for commerce and defense.[13]

Even within the single use of seashores for recreation, conflicts are inherent. Sun bathing and swimming, principal recreational uses of temperate beaches, demand clean, sandy beaches. Boating requires clean water, as do swimming and fishing. Marinas for pleasure crafts usually require the dredging of channels through flats and the filling of tidal marshes. On the other hand, hunting, another shore sport, depends on productive marshes. The aesthetic qualities of marshes appeal to bird watchers, artists, and photographers, who may or may not favor hunting. As with subsistence and commercial uses of the estuary, shoreline recreation displays inherently conflicting tendencies that confuse the supposedly distinct traditions of conservation and preservation.[14]

An ecological approach to conservation reconciles the inherent ambiguities of use and aesthetic appreciation by treating these aspects of re-

[12] Clarence Glacken, "The Origins of the Conservation Philosophy," *Journal of Soil and Water Conservation* 2 (1956): 53–56; Thomas Le Duc, "The Historiography of Conservation," *Forest History* 9 (October, 1965): 23–28; Lawrence Rakestraw, "Conservation Historiography," *Pacific Historical Review* 42 (1972): 271–81.

[13] Ranwell, *Ecology of Salt Marshes*, pp. 3–178; Davis, "Influences of Man," pp. 504–21; Willard Bascom, "Beaches," *Scientific American*, August, 1960, pp. 1–12.

[14] Roger Revelle, "Outdoor Recreation in a Hyper-Productive Society," *Daedalus* 96 (Fall, 1967): 1172–91; "Rim of the Sea Becoming Crowded and Dirty," in *Man's Control of the Environment* (Washington, D.C.: Congressional Quarterly, 1970), p. 27; John S. Banta, "Constitutional Issues and Estuarine Management," *Oceanus* 19 (Fall, 1976): 64–70; Ira Michael Heyman, "The Great 'Property Rights' Fallacy," *Cry California* 3 (Fall, 1968): 29–34.

source manipulation as interdependent.[15] Ecology provides an explanation of a particular resource's role in nature, how that role is sustained, and what conditions the resource imposes on adjacent natural resources. Also, the ecological view adds a scientific aspect to the historic and aesthetic dimensions of preservation. Rationally applying ecological principles to resource-management decisions, thus reconciling the dual trends of preservation and conservation, has led to the emergence of an estuarine preservation ideal.

An estuarine preservation ideal as part of ecological conservation is based on what wildlife biologist Aldo Leopold called the biotic health of an environment.[16] Biotic health depends on conservation measures ensuring the sustained functioning of all parts and processes of an estuary and thus guaranteeing the continued interaction of physical, chemical, and biological aspects of nature. Biotic health is synonymous with biological integrity because the protection of any environment's living creatures is assured when the integrated functions of energy conversion and transfer, nutrient recycling, water transport, and oxygen enrichment are maintained.

While the goals and policies of wetlands ecology developed gradually, it was not until a wider appreciation of new ecological perspectives spread after World War Two that national estuarine preservation became a legal possibility. The National Estuary Protection Act of 1968 committed the country to plan for the protection of ecologically sensitive coastal wetlands of unsurpassed natural value and to balance their preservation against the needs of recreation and urban or commercial development.[17]

This progression in ideas and laws from treating tidal wetlands as disdained common wastelands to recognizing them as productively rare public-interest lands occurred in three sequences involving two related stories.

One story involves the historic patterns of land and water use in estuaries, while the other is concerned with the influence of scientific ideas on cultural values and specific expressions of social concern for the environment. The principal theme in the narrative of the land is that humans re-

---

[15] S. Dillon Ripley and Helmut K. Buechner, "Ecosystem Science as a Point of Synthesis," *Daedalus* 96 (Fall, 1967): 1192–99; Odum, *Fundamentals of Ecology*, pp. 408–31, 510–16.

[16] Aldo Leopold, *A Sand County Almanac*, pp. 194–95; Nash, *American Environment*, 182–99.

[17] "National Estuary Protection Act" (PL 90-454), *16 USCA 1221*, August 3, 1968, p. 1.

shape the very contours of river mouths and significantly alter physical processes in estuaries. Because of their dynamically reciprocal character, land alterations in turn influence human assumptions, and converting a physical setting often engenders values that stimulate the creation of new resource policies. A prevailing assumption in the narrative of science is that scientific findings rarely influence policy decisions until some natural catastrophe, traceable in part to human ignorance, offers an opportunity for scientists with vision and social responsibility to suggest environmental safeguards.

As cultures convert landscapes, changing economic conditions, novel technical skills, and new resource demands create opportunities to reevaluate and reformulate laws regulating resource use. This reevaluation process affords a chance to integrate current aesthetic values and scientific evidence into the traditional policy framework. Altered conditions, advancing knowledge, and new ethics must adhere to existing legal doctrines until they together cause those doctrines to be changed.

For example, both public access to tideland fishing or hunting grounds and governmental promotion of navigation for commerce or war are legal obligations derived from ancient Roman and English jurisprudence. The judicial tradition in the first case holds that lands lying between the highest and lowest tides are imbued with a perpetual public interest, which in America is legally protected by the states under the public trust doctrine. In promoting interstate commerce, the federal government invokes the doctrine of navigational servitude, which overrides any rights granted by the states or common law to persons hindering the navigability of streams, rivers, or estuaries. Both the navigational servitude and public trust doctrines have traditionally pertained to estuarine water and land use, but recently these doctrines have been interpreted to require as well the protection of tidal marshes and water quality. Currently these doctrines are even being applied to inland rivers and wetlands as part of state and federal responsibilities for the stewardship of wildlife resources, but this interpretation is being challenged.

Changes in the use of estuarine resources may be grouped in three phases: the preindustrial, industrial, and suburban. Each phase has required a resynthesis of law, knowledge, and values because of changing demands. Although the phases overlap each other, they may be isolated for study.

During the earliest phase, the colonial, preindustrial era, the cultural dependence on the estuaries along the Atlantic shore for economic subsistence made river mouths a nexus of American maritime development. The attitudes held, techniques employed, and laws enacted by the settlers of the Atlantic tidewater would serve as precedents for the development of estuaries throughout North America. In the tidewater region, fishing and navigation dominated estuary use, with some reclamation and restriction of dumping along estuarine shores. Like all early information gathered about American geography and biology, estuarine knowledge was collected by unspecialized physicians, clergymen, or amateur naturalists. Game laws and the regulation of the wetlands were largely matters of local interest and control. Tideland fishing and hunting were accessible to most tidewater inhabitants.

As the population of the tidewater region increased and commerce grew in importance, scattered examples of reclamation occurred between Baltimore City and Boston. By the late eighteenth century, before the close of the preindustrial era, reclamation emerged as a criterion of progress. Between the late eighteenth century and the 1880s, physicians, naturalists, artists, and writers began to express an appreciation for the undeveloped marshlands as necessary natural areas, sanctuaries for the country's unique wildlife, and solitary places of rare beauty.

The second phase of estuarine development, the nineteenth-century industrial era, occurred at different times depending on the location of the particular river mouth with respect to population. The northeastern tidewater region experienced industrial influences first. By the end of the Civil War, in all the more important estuaries from the old Northeast to New Orleans and to San Francisco Bay, steam and overland transportation had developed to the point that estuaries became a terminus for industrial culture. Navigation and reclamation encouraged dumping and waste disposal in river mouths and dominated state or federal estuarine resource policies. Specialized scientists, technicians, and engineers provided the necessary data for nature study and hydraulic engineering projects.

As estuarine uses shifted from preindustrial to industrial patterns of exploitation, a corresponding change in national sentiments concerning seashores gave rise to the beginings of an estuarine preservation ideal. Steam transportation brought geographically isolated seacoasts closer to urban populations, and the popularity of travel to seaside resorts indicated a rising middle class of tourists seeking recreation and seasonal lodging in areas of rare scenic beauty. Monumental natural features especially attracted travelers, and places such as Mount Desert Island, Maine, or Cape May and Long Branch, New Jersey, received numerous visitors. A growing penchant for idyllic seaside retreats predisposed many Americans to protect certain recreational seashores.

As navigation and reclamation for commercial and residential space in urban seaports encroached on public rights to hunt and fish in tidelands, the public trust doctrine was reinterpreted by the courts and state legislatures to favor harbor improvements and development for shipping and railways. Local public health departments demanded the filling of noxious tidal flats and marshes fouled by uncontrolled refuse disposal. Federal appropriations to control siltation by dredging channels and filling tidelands were authorized on an individual basis and funded out of the general revenues.

The Swamplands Acts of 1849 and 1850 further signaled reclamation's emergence as a national priority in the use of estuaries and river valleys. The swamplands legislation relinquished the submerged areas of the federal public lands to states of the Gulf and Pacific coasts and other western states for drainage and reclamation. This split in federal and state authorities over resources quickly led to the deterioration of the nation's rivers, with losses in wildlife habitats and river and shore fisheries and with growing water pollution from sewage and industrial wastes.

After the Civil War a movement began to "regenerate" the American landscape. Inspired by concerned laymen and professional scientists, local, state, and federal governments developed policies to improve estuarine navigation, protect the wildlife of coastal wetlands, and promote the use of estuarine shores for parks. In short, an emerging estuarine preservation policy helped shift the focus from the conflicts between reclamation and navigation to the broader issues of conservation, management, and protection of natural resources. During this period visionaries and scientists fundamentally altered the national response to natural resources. They advocated multipurpose regional surveys for planning resource use, and they suggested interagency cooperation on the creation and implementation of resource policies. These suggestions are part of estuarine protection policies today.

By the 1920s, the once-dominant policies of reclamation and navigation were circumscribed by additional national policies for wildlife refuges, multipurpose river conservation, public health, and recreation. New agencies had been created to foster the specific demands of various special interests. Coordination among competing agencies was lacking; no cohesive philosophy placed their competing programs in any effective framework. The political lobbying of diverse local interests for federal funds remained the only method of deciding national resource priorities. General tax revenues were appropriated to local districts through highly specific legislative grants spent by competing federal bureaus on a case-by-case basis. Because of its organization and political experience, the Army Corps of Engineers dominated this cycle of federal patronage to local congressional districts. Specialized scientists provided the data to support each agency's conservation guidelines but offered little interdisciplinary information to gauge the influence of one program on another.

Throughout the period of Progressive conservation, the third phase of wetlands-protection history, additional scientific evidence slowly accumulated, undermining comprehensive riverine management as it had been practiced in the first four decades of this century. Before the Great Depression, the infant science of ecology had an outlook but not a methodology for distinguishing between the beneficial and costly environmental influences of single-purpose federal programs. The gradual perfection of a cross-disciplinary method for comparing scientific observations about natural environments allowed ecologists to determine theoretically the impact of competing resource policies on the long-term biological health of natural resources. This method was essential to comprehediing the role of each part of an estuary in sustaining healthy river systems. The health of any geographical setting or ecosystem, such as an estuary, is determined by the integrity of various biogeochemical cycles, which are made up of the numerous ecological connections between the living and nonliving physical parts of a landscape. Disruption of any part of a biogeochemical cycle in a natural system by a particular resource-use program affects the dominant species in the biotic community, and these effects can be measured. For example, upstream impoundment of fresh water devastates brackish oyster fisheries by allowing saltwater intrusion to replace the diverted fresh water.

Between the early 1930s and late 1940s, new quantitative analysis based on physics and biochemistry made it possible to compare different ecosystems' potentials for converting solar energy into food through photosynthesis. Ecologists calculated the energetic efficiency for sustaining life of certain natural associations like lakes, wetlands, pastures, and farmlands. Their findings were popularized by Aldo Leopold, who noted that ecological evidence showed wetlands to be remarkably efficient converters of energy into food for diverse wildlife. Leopold's reverence for land as an ecosystem that transforms solar energy into food, sustaining life's diversity and biotic integrity, was applied by Rachel Carson to the seas and their shores. Together, Leopold and Carson popularized the role of marshlands and mud flats within the estuarine ecosystem and of estuaries in the maintenance of human economic and psychic well-being. Their writings alerted the postwar public to the unintended effects of narrow scientific advice and single-purpose federal programs on the qualities of water, wildlife, and society.

In the early 1960s a cohesive estuarine preservation ideal emerged from evidence on the relation between the marshy borders and the productive potential of the estuarine environment. Experiments conducted by the Sapelo Island Research Institute of the University of Georgia revealed that tidal marshes are the unsurpassed natural food-producing areas of the estuarine complex. Tons of food produced by the marsh grasses, algae, and diatoms are pumped throughout the estuarine system by the tides, adding to the productive capacity of estuaries to sustain land and sea life. Flooding rivers frequently flush these nutrients out to sea, contributing to the productivity of near-shore fisheries. The long-held suspicion that marshes may be chemically necessary for sustaining plant and animal life was thus clarified and confirmed by quantitative ecological analysis.

As ecologists recognized estuaries as a weak spot in Progressive concepts of comprehensive riverine conservation, suburban expansion with new industrial land and water uses began impinging on estuaries and coastal regions. Both through suburban shoreline developments and through the accessibility of households to seashore recreation, the widespread use of the automobile contributed to growing pressures on coastal areas. The suburban transformation also placed growing demands for electricity and clean water on already exploited estuarine areas. Electrical generating plants discharged heated water into rivers and estuaries, while petrochemical plants added to coastal water pollution. The competition between these uses and suburban demands for residences and recreation facilities ultimately destroyed the old Progressive consensus concerning the allocation of scarce estuarine resources.

New scientific evidence, accelerating social demand, and limited coastal resources profoundly influenced conservation legislation in the 1960s. The search for a workable estuarine preservation policy by the states during the 1950s produced various federal wetlands legislation, culminating in the National Estuary Protection Act (1968). That legislation brought together political strategies and planning techniques from the early New Deal. Interbureau coordination and comprehensive resource decisions were promoted by agreements between the secretary of the Interior and the secretary of the Army concerning the effects of dredging, dumping, and reclamation in estuaries. The law established several national precedents in environmental policy review and in requiring surveys to identify critically important ecological marshlands for protection. Review provisions of the National Estuary Protection Act were later incorporated into both the National Environmental Policy Act, 1969, and the Coastal Zone Management Act, 1972. The estuary protection legislation required an initial assessment of the effects of estuarine dredge or fill requests on the water quality and wildlife of the ecosystem. The act encouraged protection of wildlife and water quality through local and state planning agencies' cooperation with federal authorities to identify tidal marshes and to plan comprehensively for their perpetual preservation as open space.

Despite the strides made by the 1968 act, problems in estuarine protection remain. The incompatibility of resource uses, such as the preservation of tidelands versus their reclamation, is still compounded by conflicting governmental jurisdictions.[18] While the ownership of land within estuarine areas is complicated, the water and wildlife are considered the common property of all the people and are variously protected by state and federal laws. The Army Corps of Engineers has responsibility for the maintenance of navigation channels, while the tidelands are held in trust by various coastal states to enhance public access to the shore. Some tension exists in the states' guarantee of public access to the tidelands, since navigation, fishing, and hunting are all subsumed under this doctrine. Local cities or other districts have repeatedly lobbied both state and federal authorities to reclaim entirely certain tidal marshes for the purposes of agriculture, transportation, or housing. There is competition even among agencies at the federal level. The U.S. Fish and Wildlife Service, for example, administers wildlife refuges in tidal wetlands, but the Bureau of Reclamation, among others, has encouraged the impoundment of instream water flow by dams, depriving the marshes and the refuges of life-sustaining water and sand.

[18]Uzes, *Chaining the Land*, pp. 123–43; Aaron L. Shalowitz, *Shore and Sea Boundaries*, I, 89, 318, and II, 530–41, 612, 675–79.

These conflicts will no doubt prove intractable because many incompatibilities among local, state, and federal policies for conservation merely reflect the problems inherent in the multifaceted economic uses of the shore.

These problems notwithstanding, a keener understanding spreads. Ecologists have provided a framework to unify the aesthetic, scientific, and economic arguments underlying environmental protection as conceived in an estuarine preservation ideal. It is now understood that the natural features of estuaries form a tightly integrated system for capturing solar energy and converting a portion of that radiation into food for bacteria, shrimp, fish, insects, birds, and mammals. Estuaries are very efficient energy converters because of their diverse natural features. Oyster reefs, mud flats, marshes, channels, sand bars, and beaches constitute a cohesive natural association in river mouths or along coasts. Marine ecologists classify each of those natural features as a distinct habitat supporting a discrete biotic community. Geographical habitats, the non-living part, and biotic communities, the living or once-living aspects of any environment, comprise an ecosystem. The components of ecosystems geologically and biologically coevolve. This fact enables ecologists to define existing relations among any population and its adjacent territory over time and to predict the probable consequences of environmental changes for various creatures, communities, and habitats.[19]

The predictive ability of ecology to determine the relative influences of resource extraction on biotic communities makes possible the formulation of conservation policies enhancing the long-term health of the environment and the sustained yield of vital resources. Aldo Leopold referred to the goal of these conservation policies as stewardship of natural resources, which he believed required that the environment's biotic health further the coevolved diversity of living things. Coevolved diversity results from the natural interdependence among soils, plants, and animals and enables life forms to adapt successfully to changing environmental conditions.[20]

Coevolved diversity means that the continued existence of blue herons or harbor seals in estuaries depends now, as in the past, on the maintenance

[19]Odum, *Fundamentals of Ecology*, pp. 5, 8–23, 251–64, 352–61, 405–507, 510; Miller, *Living in the Environment*, pp. 43–88; Donald Worster, *Nature's Economy: The Roots of Ecology*, pp. 300–306; Ripley and Buechner, "Ecosystem Science," pp. 1192–99.
[20]Leopold, *Sand County Almanac*, pp. 251–64.

of estuarine conditions that make possible the varied evolution of these birds' and mammals' diverse prey. To the ecologist, life's evolution, diversification, and success are as much communal efforts as individual genetic achievements. All creatures coevolve because of the intimately shared dependence of predator and prey alike on common territories. Stewardship is theoretically attainable by reasonably allocating resources to protect estuaries from sudden, stressful, or shocking changes that may disrupt the life cycles of critical species of prey. Leopold suggested that such stewardship assured the protection of aesthetic values as well, because biological health or integrity is a sustaining factor in what humans find beautiful in nature.[21]

Still, the public interest in tidal seas is so ancient and so multifaceted that competition for the use of estuarine resources is inevitable. Indeed, the repeated redefining of the biblical concept of man as God's steward by medieval and modern cultures is a primary reason for the tensions in use, protection, and appreciation of natural resources, according to environmental philosopher John Passmore. Even today stewardship is defined differently by various competing constituencies. Conservation as stewardship provides the legal and administrative means to assure equitable, responsible, and efficient distribution of these resources. As part of conservation, an estuarine preservation ideal and the methodology of ecological science provide guides to the probable consequences of specific conservation actions on the ecological functioning of river mouths and afford a coherent view of interdependence of estuarine ecosystems and a goal of preservation of complexity and diversity within those systems.[22]

Throughout the nation today a series of state and federal estuarine refuges exist as quiet testimony to the ideals, efforts, and commitment of local conservation groups, planners, engineers, and scientists. These advocates possess a resolute maturity in asserting that some places must be set aside for future generations because, as Rachel Carson once remarked, "man's way is not always the best."[23]

[21] N. S. Shaler, *Man and the Earth*, p. 93; Leopold, *Sand County Almanac*, pp. 280–95.

[22] John Passmore, *Man's Responsibility for Nature: Ecological Problems and Western Traditions*, pp. 28–40, 73–126, 173–95.

[23] Paul Brooks, *The House of Life: Rachel Carson at Work*, p. 226.

2

A Frontier of Estuaries

A bolder and safer coast is not known in the universe:
to which conveniences, there's the addition of good
anchorage all along upon it.

—Robert Beverly, 1705

THE river valleys of the Atlantic shore cut through the coastal plain, creating large numbers of estuaries, many of which are bordered by extensive tidal marshes. These coastal wetlands have been held in public trust since the original colonial grants. Unlike rights in other portions of the public domain, the public rights to fishing, hunting, and navigation on tidal lands could not be extinguished by sale to private concerns. Thus the general trend of granting land in fee simple absolute was not followed in the Atlantic, Gulf, and Pacific tidelands.[1]

Seawater along the coast meanders upriver to the farthest reaches of the tidal flood. The back-and-forth motion of the river flow and tides is a unique feature of the estuarine portions of the river. The tidewater district—that portion of the lands along the Atlantic Coast encompassing these estuaries—developed as the earliest location of European settlement. Two centuries after Ponce de León initially landed on the Florida shore, the Atlantic Coast remained the center of colonial culture. During the three centuries separating the founding of Saint Augustine in 1565 and the American occupation of the Pacific Coast, the tidewater was the focus of

---

[1]U.S. War Department, Army Corps of Engineers, *Shore Control and Port Administration: Investigation of the Status of National, State, and Municipal Authority Over Port Affairs*, pp. 25–34. Tidelands, defined as those areas of the coast ordinarily exposed by daily tides over a nineteen-year period, often correspond to the biological classification of the littoral zone (See Aaron L. Shalowitz, *Shore and Sea Boundaries*, I, 89, 318, and II, 530–41, 612, 675–79).

national development.[2] Consequently, tidewater landscapes—having borne the burden of population expansion, technological change, and urban growth—have been transformed more thoroughly than any other region.

The oldest human economy sustained by estuarine resources was fishing and hunting. Since the retreat of the last or Wisconsin glaciation, the advancing sea level has submerged many archaeological sites along ancient shorelines. Underwater archaeology has discovered evidence of human settlements in southern California in the third millenium B.C. An archaic, maritime-dependent culture has been traced to the second millenium B.C. in Santa Barbara, and a fish weir unearthed in Boston has been dated to 1700 B.C.[3] Fisheries encouraged dense Native-American coastal populations before prolonged European-borne epidemics decimated them in the sixteenth century.

In the coastal areas, fish, shellfish, and seaweeds were used as food by indigenous peoples. The marshes provided reeds for housing and ducks, geese, and swans for food. Muskrat, mink, and raccoon hides were also obtained. Tidelands provided Native-American nations that seasonally hunted in the marshes the raw materials for food, clothing, shelter, and tools. For the indigenous peoples, unlike their European dispossessors, marshlands were important sources of raw materials and not places to live.[4] However, during the times of continuing resistance to white invasions, many Natives took refuge in swamps—from the Pequots and the

[2]Merle E. Curti, *The Growth of American Thought*, pp. 31–33, 44, 53, 64; Ralph H. Brown, *Historical Geography of the United States*, pp. 3–5, 7–10, 19–23, 43–67, 68–75, 89, 98; Ralph H. Brown, "The Land and Sea: Their Larger Traits," *Annals of the Association of American Geographers* 41 (September, 1951): 199–209; Harold Kirker, *California's Architectural Frontier*, pp. vii–viii; Earl Pomeroy, *The Pacific Slope*, pp. 3, 5–9, 14, 15–19, 31–33. *Tidewater* and *seaboard* are synonymous and designate regions whose seaward boundary is defined by the tidelands of the shore.

[3]Harold Driver, *Indians of North America*, pp. 3–7, 56–60, 98–100, 213–14, 221, 323–24, 563; *Atlas of United States History* (New York: American Heritage Press, 1966), pp. 18–19; Frederick Johnson et al., *The Boylston Street Fishweir*, Papers of the Robert S. Peabody Foundation for Archaeology (Andover, Mass.: Phillips Academy, 1942), pp. 24–38, 131–48; Martin Baumhoff, *Ecological Determinants of Aboriginal California Populations*, American Archaeology and Ethnology No. 49 (1958–1963), pp. 155–223; Preston Cloud, *Cosmos, Earth, and Man: A Short History of the Universe* (New Haven: Yale University Press, 1978), p. 266.

[4]Francis Jennings, *The Invasion of America: Indians, Colonialism, and the Cant of Conquest* (New York: W. W. Norton, 1975), pp. 26, 255, 261–62, 282–362; Baumhoff, *Ecological Determinants of Aboriginal California Populations*, pp. 155–223.

Wampanoags of New England to the Creeks and the Seminoles of the Southeast.

No single pattern of ritual existed among Native Americans to describe their diverse attitudes toward water, fisheries, or man's place in nature. However, all of these societies regulated the knowledge and use of certain fishing and hunting areas by inheritance. For many native cultures, their surrounding landscape was a historical point of reference from which succeeding generations derived an identity. Many Native Americans thereby transcended a merely economic relationship to the land.[5]

The earliest documented European landing on any North American coast occurred at a river mouth in Newfoundland, the Beothuk tribal area, during the year A.D. 1001. Historian Samuel Eliot Morison has commented that Lief Ericson's landfall happened in an area where the salmon were large and profuse.[6] Thus, the North American fisheries were "discovered." It was local declines in fisheries that later lured Basque whalers and Breton fishermen farther from their native shores. By the 1480s these mariners had ventured to the Grand and George's banks off North America. So crucial would American fisheries become to European governments that after the French lost the Seven Years War, they managed to retain fishing islands in the Gulf of Saint Lawrence even while they forfeited an overseas empire.[7]

The first known landing of a southern European in the continental United States occurred along the east coast of Florida with Juan Ponce de León's arrival during the first week in April, 1513. While searching for a fabled fountain of youth, the Spanish expedition met with Native-American resistance, and Ponce de León returned to Puerto Rico with only 170 turtles, fourteen seals, numerous sea birds with their eggs, and charts of the Florida coast from the Saint John's estuary to the Coral Keys.

The most significant estuaries of the Atlantic Coast were detailed in 1524 by the Florentine pilot Giovanni da Verrazano, who charted the re-

[5] Frank G. Speck, "Land Ownership among Hunting Peoples in Primitive America and the World's Marginal Areas," *Proceedings of the 22nd International Congress of Americanists* (Rome, 1926), II, 323–32; Erik Erikson, *Family, Childhood and Society*, 2nd ed. (New York: W. W. Norton & Co., 1963), pp. 166–70; A. L. Kroeber, "The Yurok," *Handbook of California Indians*, Bureau of American Ethnology Bulletin 68 (Washington, D.C.: Government Printing Office, 1925); Charles Bowden, *Killing the Hidden Waters*, p. 8.

[6] Samuel Eliot Morison, *The European Discovery of America: The Northern Voyages*, p. 46.

[7] Sir Allister Hardy, *The Open Sea: Its Natural History*, pp. 1–67, 215–32, 247–93; Carl Ortwin Sauer, *Sixteenth Century North America*, pp. 3–76, 154–56; R. E. Coker, *This Great and Wide Sea: An Introduction to Oceanography and Marine Biology*, pp. 260–62.

gion between Cape Hatteras and Nova Scotia. He found New York Bay particularly "commodious and delightful." [8]

Arthur Barlowe, the first Englishman to portray the natural resources of the Atlantic Coast, substantiated one of the most useful aspects of the estuaries. In 1584 he described the fishery wealth of Pamlico Sound, North Carolina, where he had watched a Native-American fisherman: "In less than half an hour he had laden his boat as deep as it could swim." Barlowe's expedition also returned with the first English watercolors of the American seashore. The drawings of John White, commissioned by Sir Walter Raleigh, vividly depict the peoples, vegetation, and wildlife nestling among the marshes of Roanoke's barrier island. [9] One particular painting labeled "The Manner of Their Fishing" bears a remarkable stylistic affinity to Sandro Botticelli's "Birth of Venus" (1480) in its unrealistic portrayal of the sea. The sea lacks movement, perspective, and aquamarine opacity. In White's painting, he attempted to show a variety of marine life in addition to three methods of native fishing—from a canoe, spear fishing on mud flats, and with a fish weir in shallow waters.

In 1609 Henry Hudson's expedition noted "many salmons, mullets, rays, very great" in New York Bay. Later, David de Vries was more precise as to native customs, commenting that "striped bass are caught in large quantities and dried by the Indians." He described fish weirs of sticks and netting used along the littorals to corral fish at the flood tide so as to strand them at the ebb tide. No edible item, from the tiny limpets and periwinkles grazing on the algae-covered rocks to the extensive oyster banks thriving in the less saline portions of the river's mouth, was overlooked in this subsistence economy. Even fishheads were used, as fertilizer to renew the plots of squashes, beans, and maize that supplemented the Atlantic aboriginal diets. The early colonial European settlements relied equally on the bounty from the sea. [10]

During the seventeenth century, Captain John Smith described the lands surrounding North America's largest estuary, the Chesapeake Bay. He portrayed the region as "watered so conveniently with fresh brooks and

[8] John Bakeless, *The Eyes of Discovery*, p. 203. See also John Conron, *The American Landscape*, pp. 99, 105–106.

[9] Richard Hakluyt, *Hakluyt's Voyages: The Principal Navigations, Voyages, Traffiques, and Discoveries of the English Nation*, ed. Irwin R. Blacker, p. 288; Sauer, *Sixteenth Century North America*, pp. 261–64, 281.

[10] Bakeless, *Eyes of Discovery*, pp. 184, 196–97, 218–24, 238; Alanson Skinner, *The Indians of Greater New York*, pp. 42–43.

springs, no less commodious than delightful." European sentiments about estuaries and tidal marshes were influenced by this repeated characterization of the environs as both useful and beautiful by Renaissance explorers.[11]

Smith was a product of the optimism, utility, perseverance, and opportunity that characterized his era. A promoter of the Atlantic shore, he wrote, "No place is more convenient for pleasure, profit, and man's sustenance" because, he noted, "the waters, isles and shoals are full of safe harbors for . . . boats of all sorts," including both transportation and fishing. For Smith and some later preindustrial residents, the Chesapeake's biotic resources complemented the benefits afforded by harbors.[12]

On the other hand, later travelers in the Chesapeake region were quick to point out the adverse conditions of early settlements. Captain Nathaniel Butler, writing of Virginia in 1622, "found the Plantacons generally seated upon meer Salt Marishes full of infectious Boggs and muddy Creeks and Lakes." Consequently, Butler concluded that the Jamestown settlement was "subjected to all those Inconvenyenceies and diseases, which are soe commonly found in the most unsound and most unhealthie partes of England."[13]

Narratives subsequent to de Vries's, Smith's, and Butler's exhibited ambivalent feelings shared by colonials about their tidewater environment. Productive potentials of the estuaries were exploited, while the adverse qualities were avoided, remedied, or at least decried. Robert Beverly's description of Virginia at the opening of the eighteenth century discussed at some length the amenities and drawbacks of Chesapeake Bay life. Beyond extolling the ease of coastal navigation, he observed, "Springs flow so plentifully, that they make the river water fresh . . . sometimes a hundred miles below the flux and reflux of the tides." Seasonal infestations of shipworms were mentioned as a serious threat to shipping: "In the month of June annually, there rise up in the salts, vast beds of seedling-worms which enter the ships, sloops and boats . . . and eat the plank . . . like those of a honeycomb." He advised four methods of avoiding the infestation, the last

[11] J. Thomas Scharf, *The Chronicles of Baltimore: Being a Complete History of "Baltimore Town" and Baltimore City from the Earliest Period to the Present Time*, pp. 2–3; John Lankford, ed., *Captain John Smith's America*, pp. 4–5, 7.

[12] John Smith, *Arguments for Colonization* (1614), quoted in Phillip Viereck, *The New Land*, p. 17.

[13] Michael Kammen, *People of Paradox: An Inquiry Concerning the Origins of American Civilization*, p. 149; Phillips, *Life and Labor in the Old South*, p. 52.

of which was "by running up into the freshes." Beverly explained that "they never bite nor do any damage in fresh water or where it is not very salt." [14]

Beverly noted that rice, corn, and hemp could be grown in the heavier soils near the river's mouth, while submerged sandy tracts supported cranberries or huckleberries. "The rivers and creeks . . . in many places form very fine large marshes," he wrote, "which are a convenient support for . . . flocks and herds." The tidal marshes also afforded wild grapes, wax myrtle for candle making, and large numbers of beaver, otter, mink, and muskrat. But Beverly was most impressed by the fish: "Both of fresh and salt water, of shellfish and others, no country can boast of more variety, greater plenty or of better." Spring and summer brought the migration of herring, followed by shad, sturgeon, trout, and bass. He concluded that "as in summer, the rivers and creeks are filled with fish, so in winter they are in many places covered with fowl." Like other observers, he exclaimed, "The plenty of them is incredible." Both the subsistence farmer and the planter were well-nourished, for as Beverly noted, "the shores, marshy grounds, swamps and savannas are also stored with like plenty of other game." [15]

Southern tidewater historians reveal that the Chesapeake marshes, unlike the Carolina rice districts, remained virtually unreclaimed throughout the eighteenth century. Despite an act passed by the House of Burgesses in 1712 to encourage the drainage and cultivation of marshes, swamps, or overflowed acreage, the Virginia aristocracy continued to utilize these grasslands for pasturage. Despite the need for newly arable lands required by tobacco planting, custom dictated that inferior grades of tobacco, requiring longer aging, be produced on marsh or bog soils. Although larger plants were obtained from drained soils, tobacco economy encouraged the clearing of woodlands for new fields rather than the drainage of swamps. Similarly, until the mid-eighteenth century, South Carolina's rice culture had remained along upriver marshes; only later did it move to the tidal marshes near the river's mouth. Contrary to the northern experience, colonial expansion in the Virginia and Carolina tidewater had less effect on the tidal marshes than later agrarian seashore cultures would. [16]

[14] Robert Beverley, *The History of the Present State of Virginia*, ed. Louis B. Wright, pp. 121–27.

[15] Ibid., pp. 146–49, 153.

[16] Phillip A. Bruce, *Economic History of Virginia in the Seventeenth Century*, pp. 430–35.

The colonial Europeans created imperial outposts and were not genuinely tied to an indigenous economy. Despite John Smith's encouraging words about the Chesapeake Bay's resources, settlements along the Virginia and Maryland tidewater turned to the cultivation of native tobacco for export to European markets. Even in coastal New England, where fortunes were made from the cod fishery, the major impetus was to support a triangular trade in slaves and sugar.[17]

The Europeans who settled along the northeastern littoral excelled in navigation and ocean commerce, and their dependence on seaborne trade had important consequences for the landscapes of growing settlements. Early in its development Boston merchants petitioned the Bay Colony to allow wharf construction out into the Town Cove waters. On November 29, 1641, the town council granted the present site of Faneuil Hall Square to a merchant association in order to convert the marshland into wharves and warehouses. Two years later a northern cove and its surrounding marshes were granted to several entrepreneurs, provided they "erect and make upon or neere some part of the premises, one or more corn mills. . . ." By 1645 fifteen private wharves had been constructed in what was then the largest of the colonial port towns.[18]

On his return to Boston, in 1663, John Josselyn commented, "The houses are for the most part raised on the Seabanks and wharfed out at great industry and cost, many of them standing on piles, close together on each side of the streets as in London." Another major accomplishment for the promotion of commerce was the construction of the Long Wharf through the Town Cove and well into the harbor in 1710. This area remained free from further filling until 1872, when Fort Hill was razed to provide the necessary fill. Thus, while the most extensive reclamation around Boston was to occur in the nineteenth century, the human alteration of the Shawmut peninsula had early beginnings.[19]

An immediate assumption of communal responsibility for the environment was also demonstrated in 1636 by the appointment of "water bailiffs . . . to see that noe annoying things be left or layd about the seashore." Eleven years later the idea of public access to the shore was guar-

---

[17] Samuel Eliot Morison, *A Maritime History of Massachusetts*, pp. 12–19; William Brandon, *The Last Americans* (New York: American Heritage Press, 1961), p. 166; Joseph M. Petulla, *American Environmental History*, pp. 29–47.

[18] Walter Muir Whitehill, *Boston: A Topographical History*, pp. 4–12.

[19] Ibid.

anteed by the General Court of the Massachusetts Bay Colony. The statutes referring to "common liberties" established that "everie . . . householder shall have free fishing and fowling in any great Ponds, Bays, Coves, and Rivers so far as the Sea ebbs and flows, . . . unless the Free-men of the same town, or the General Court have otherwise appropriated them." [20] While attempting to strike a balance, the ordinance ensured that the immediate interests of a few would prevail over the long-term benefits held in common by the free citizens. In this way, dual use of the estuary was codified in law, and provision made that should these uses ever mutually exclude each other hunting and fishing would take a secondary role to commerce.

As the coastal settlements grew into prosperous commercial centers, the filling of tidal marshes furthered the maritime success of early American civilization. The development of Baltimore is a case in point. Maryland's tidelands were granted by King Charles I to the Lord Proprietor, who in turn had the power to dispose of such lands, subject only to the public rights of fishing and navigation. Beginning in 1683, commercial public rights were granted to landowners on the Patapsco River. However, it was not until 1729 that orders "for erecting a town . . . and for laying out lots on sixty acres of land" were passed by the Colonial Assembly. These directives were carried out in 1730, and thus Baltimore City "fixed itself amid the creeks and marshes and under the hills of the northwestern branch of the Patapsco." An assembly act of 1745 stated that any kind of improvement made along watercourses would belong to the improver in perpetuity. Such action was necessary because Baltimore's extensive marshes obstructed commercial access to the shore by all but one street. In 1750 the first public wharf was constructed, attesting to the growing importance of Baltimore's waterborne trade. The concern over the navigability of the waterways was demonstrated by the passage of an antidumping measure in 1753. After years of tobacco culture, the increase in siltation resulting from the extensive removal of ground cover required legislation in order to protect navigation "below the High water mark." [21]

[20] Carl Bridenbaugh, *Cities in the Wilderness: The First Century of Urban Life in America, 1625–1742*, p. 24; *Book of the General Laws and Libertyes of Massachusetts* (1636), ed. Thomas G. Barnes (facs. ed.; San Marino: Huntington Library, 1975), p. 35; Morison, *Maritime History of Massachusetts*, pp. 10–25.

[21] *Kerpleman* v. *Board of Public Works of Maryland*, 261 *Maryland Reporter*, p. 436, and 276 *Atlantic Reporter*, 2nd ed., p. 56; Scharf, *Chronicles of Baltimore*, pp. 14, 18, 35–36, 49.

Similar problems were faced by the conquerors of the Dutch colony of New Amsterdam. The Reverend Jonas Michelius described the settlement not long after Peter Minuit received rights of occupation from the Lenni Lenape (Delaware Indians) in 1626. "The country produces many species of good things which greatly serve to ease life," Michelius wrote, "fish, birds, game and groves, oysters, tree fruits, fruits of the earth, medicinal herbs and others of all kinds." On May 23, 1654, Niasius de Sille wrote a friend in the Hague that "the rivers are full of good edible fish . . . [and] oysters . . . so large that they must be cut in two or three pieces." He specifically mentioned that the surrounding waters had "perch, sturgeon, bass, herring, mackeral, weakfish, stone bream, eel and various other kinds. . . ."[22]

Lower Manhattan was also the site of the largest reclamation of marsh that took place in the seventeenth century. After 1664, an inlet of the East River called the "great Graft" was increasingly littered with filth, making the abutments along its shores useless. Today's present Broad Street was created by filling the inlet after 1675.[23] Like her European namesake, this Dutch colony turned to public assessment to cover the cost of the operation rather than encouraging private corporations to reclaim the shore.

Early reclamation as practiced in Boston, Baltimore, and New York created preindustrial urban landscapes along estuarine shores. Although these landscapes differed from the agrarian settings established on the Virginia tidelands or by the diking of wetlands in South Carolina, all of these new settings were believed by colonials to be "improvements" of the existing environment. Reclamation appeared especially necessary along northeastern shores because early waste disposal and siltation threatened commerce and public health, two endeavors with widespread constituencies.

The English jurist Mathew Hale expounded the preindustrial European attitude toward estuaries when, in 1677, he wrote that "man has a duty to protect the world from ponding of water in marsh and bog. . . ." The usefulness of such undertakings was self-evident to the post-Newtonian world of European science and mechanics. Despite the labor-intensive requirements of eighteenth-century reclamation, Count Buffon in 1749 equated the continued existence of marshes with a lack of human industry.[24]

[22] John Kiernan, *A Natural History of New York City*, pp. 4, 6.

[23] Bridenbaugh, *Cities in the Wilderness*, p. 20.

[24] Clarence Glacken, *Traces on the Rhodian Shore*, pp. 476, 479–81, 680. Glacken comments: "One of the great landscape changes in Modern times has been marsh, bog, marine, and lacustrine drainage. This extensive drainage of marsh and bog lands, however,

The underlying factor most responsible for changing the estuarine landscape was the colonial Europeans' unequalled ability to tap the surrounding environment for mechanical power. Broadly speaking, the French, Dutch, and English colonials harnessed the winds. These colonials imported diking, drainage, and mechanical wind engines from Europe. Since the eleventh century, polder draining had been practiced by the Dutch throughout Europe, and in the following century the English had constructed the first windmills. It was not until the techniques of windmill design, pumping, and dredging were combined in the fifteenth century, though, that truly grand designs upon the landscape could flourish.

By 1645 the first ladel dredger had been used in Holland for reclamation purposes. A century later French designers added a treadmill device, rather than the English winches that had been used originally to lift the ladels. These improvisations allowed Englishmen Herbert Boulton and James Watt to attach a four-horsepower steam engine to a ladel dredger at Sunderland Harbor in 1794. Such experiments fundamentally changed the capacity of society to alter the shore's contours.[25]

The success of both the urban and rural reclamation undertaken by the Dutch, French, and English demonstrates repeated dependence on technology to overcome natural hazards. The windmills brought by the colonists from Kent, Flanders, Acquitaine, and Iberia ground flour and, more importantly, pumped water from low-lying or seasonally flooded districts. Tidal grist mills were constructed to utilize the power inherent in the daily rise and fall of estuarine waters, particularly in severely constricted inlets of considerable expanse at high water. Early reclamation of estuarine landscapes was thus made possible by the ingenious use of mechanical wind machines and the application of hydraulic engineering principles to land drainage, changing the very shapes of our tidal shores.[26]

Opposition to European reclamation usually came from local fishermen and fowlers, and this opposition often delayed but never thwarted the

---

seems primarily to be a phenomenon of modern times, mostly since the late seventeenth century—although there are many . . . examples from earlier periods" (p. 348).

[25] Audrey M. Lambert, *The Making of the Dutch Landscape: An Historical Geography of the Netherlands*, pp. 3, 30–31, 45, 77–105; Johan Van Veen, *Dredge, Drain and Reclaim: The Art of a Nation*, p. 46; L. E. Harris, "Land Drainage and Reclamation," III, 303–19, G. Doorman, "Dredging," IV, 630, 641, and J. Allen, "Hydraulic Engineering," V, 522–23, all in *History of Technology*, ed. Joseph Singer; Norman Smith, *Man and Water: A History of Hydro-Technology*, pp. 29–44.

[26] Smith, *Man and Water*, pp. 29–31.

attempts of engineers to "lay dry" the land. To land-hungry European nations, reclaiming submerged tidelands and overflowed areas was the primary means to increase pasturage, extend agriculture, extract energy-rich peat, and expand urban commercial or residential space. As fuel shortages tapped the vast peat beds, stripping them of their cover, the danger from flooding increased the need for further diking and pumping.[27] Before Holland and Britain were able to dominate the seas, they had first to encounter the estuary, understand its tidal fluxes, and master the migrating sands of its shoals. With the creation of the Royal Society in 1662, accurate reclamation information was publicly dispersed and made available to the English. Atlantic colonials were eager to apply this knowledge to their own problems with reclamation.

The seacoast swamps of the Atlantic shore dwarfed any comparable areas in Europe. As long as the technology necessary for drainage of large areas of marsh remained poorly developed, their alteration proved impossible. Despite this handicap, the Delaware colonials had, as early as 1762, chartered the Saint George's Marsh Company to undertake the diking of less than 17.5 square miles of tidal marsh near Delaware City, a task first begun by the original Dutch immigrants.[28]

Between 1766 and 1767 landed Quaker merchant Thomas Gilpin agitated for the construction of a canal based on European engineering techniques to link the Chesapeake and Delaware bays. As a member of the American Philosophical Society—the colonial equivalent of the Royal Society—Gilpin presented his idea in 1769. Submitted to two committees within the society, Gilpin's scheme elicited an investment of £140 sterling from Philadelphia's merchants to finance a preliminary survey. In 1771, the society assessed the results of five surveys. The report recommended the construction of a barge canal because it would be cheaper than a ship canal. While Pennsylvanians were enthusiastic about such a proposal, Maryland feared that the Philadelphians were attempting to lure the lucrative Susquehanna River Valley trade away from Baltimore.[29]

In the long run, lack of scientific expertise and capital appear to have been the major obstacles to construction of the canal. The foremost Ameri-

[27] Lambert, *Making of the Dutch Landscape*, pp. 208–20.

[28] U.S. Department of Agriculture, Office of Experiment Stations, *Tidal Marshes and Their Reclamation*, prepared by George M. Warren, No. 240, pp. 30–34.

[29] Brooke Hindle, *The Pursuit of Science in Revolutionary America*, pp. 210–11, 212; Glacken, *Traces on the Rhodian Shore*.

can scientist of that era, Benjamin Franklin, had argued to the surveyors that some experience was necessary before any accurate cost estimates could be made and that expertise was sorely lacking in America. In a report to the colonial legislature, the two surveyors recommended importing European assistance.[30]

The difficulties encountered by Gilpin in 1766–67 in the construction of the Chesapeake and Delaware Canal were not typical of other eighteenth-century canal ventures. Three canals connecting various portions of the seaboard were forerunners of larger inland canals. Encouraged by George Washington, the earliest of these was the Dismal Swamp Canal, which connected the Virginia and North Carolina tidewater districts. In 1792 South Carolina investors promoted the connection of the Santee River to the port of Charleston by a twenty-mile-long tidewater canal, which was completed in 1800.

In 1793 Boston merchants invested in a similar venture to link their harbor with Lowell, Massachusetts, on the Merrimack River. The Middlesex Canal spanning the thirty miles between Boston and Lowell opened in 1804. Larger than the Dismal Swamp project, these two later canals were part of numerous inland navigation improvements in the late eighteenth century. Usually the ambitiousness of these schemes exceeded the mechanical sophistication of colonial technology. The attempts to link the Potomac River to Fort Pitt or the Delaware River to Lake Ontario were financially impracticable when compared with the return on a similar investment in coastal canal ventures.[31]

Canal building represented the dominant activity of eighteenth-century Americans in tidewater navigation. This understandable urge to bring raw materials to ports as cheaply as possible could be realized only with the development of hydraulic engineering. The building of locks, bridges, levees, and pumps required experimentation and technical specialization to meet the differing conditions of each particular region traversed by the canals. The drainage of swamps, like the digging and dredging of canals, was widely considered an improvement over the tidewater landscape's inhospitable features.

The last half of the eighteenth century was an important era in the formulation of American attitudes toward nature, the environment, and the

[30] Hindle, *Pursuit of Science*, pp. 210–11, 212.
[31] Brown, *Historical Geography*, pp. 103–107.

tidal marshes. Under the influence of the European Enlightenment and the growing Romantic movement, Atlantic colonial savants inherited conflicting views toward nature. This conflict was added to an older colonial ambivalence, codified in seventeenth-century laws. Yet the dominant strain within the tiny American scientific and intellectual community was Baconian and was illustrated in Benjamin Franklin's career. Francis Bacon had stressed the importance of science made practical through application.[32]

Dr. Benjamin Rush, in dealing with Philadelphia's yellow fever epidemic of 1793, exhibited his belief in practical science. As a rapidly growing population spread out around the colonial port towns, obtaining an adequate fresh water supply became a serious problem. Typhoid and yellow fever epidemics were periodic in the crowded colonial waterfronts. Philadelphia's 1793 yellow fever epidemic was an extreme example of that city's increasing "unhealthiness," mistakenly attributed by Rush to the city's vast marshes. His and subsequent addresses on the subject to the American Philosophical Society document the developing American attitude toward tidal marshes, coastal swamps, and mud flats. Rush argued that fevers "on the shores of the Susquehanna have kept an exact pace with the . . . propagation of marsh effluvia, by cutting down the wood which formerly grew in the neighborhood."[33]

The experiments of Joseph Priestly in 1772 and those of Jan Ingenhousz in 1778 were known to the small circle of colonial fellows interested in quantitative description and detailed observation of natural phenomena. These experiments showed that plants were the purifiers of the air. Therefore, in Rush's mind, removing forest cover without replanting crops adversely influenced the "economy of nature," terminology popularized by the Swedish systematic botanist Carl von Linne or Linnaeus in his 1791 discourse on natural behavior, *Oeconomia Naturea*. Rush believed that human health depended on the well-being of the surrounding environment. Without specifying his experimental proof, Rush concluded that "draining swamps, destroying weeds, burning brush, and exhaling the unwholesome or superfluous moisture of the earth, by means of frequent crops of grain, grasses and vegetables of all kinds, renders it healthy." In the following

---

[32] Hans J. Huth, *Nature and the American: Three Centuries of Changing Attitudes*, pp. 16–53; Roderick Nash, *Wilderness and the American Mind*, pp. 8–22, 44–66; Glacken, *Traces on the Rhodian Shore*, pp. 471–84; Hindle, *Pursuit of Science*, p. 190.

[33] Benjamin Rush, "An Inquiry into the Cause of the Increase of Billious and Intermitting Feveers, in Pennsylvania, with Hints for Preventing Them," *Transactions of the American Philosophical Society* 2 (1786): 205–209.

decade, the Reverend Thomas Malthus also insisted on the unhealthiness of marshlands.[34]

The American Philosophical Society dedicated several papers to the general theme of the insalubrity of marshes. In November, 1794, Thomas Wright, a surgeon, suggested the impracticality of draining extensive coastal swamps. Instead he advocated allowing more breezes to carry off "bad" marsh air by clearing a 200-mile-long corridor in the forests to funnel healthy winds toward the swamps. The following October, William Currie agreed with Wright on the obstacles to extensive drainage but differed with him regarding the best means to render coastal swamps innocuous. Larger marshes, he suggested, "should be constantly flooded by means of dams and sluices to prevent the effects of putrefaction." Currie also challenged the persistent notion that seasonal diseases were caused by "invisible exhalations or miasma" from marshes. Instead he argued that decaying vegetable and animal matter, when exposed to the air, carried disease and foul odors.[35]

Of all the reported experiments that the Philadelphia epidemics sparked, none were more significant than those performed by Dr. Adam Seybert. His findings, read before the society on December 21, 1798, demonstrated that "the putrefaction of the animal and vegetable matters upon the soil of marshes, was the great cause of" the recurrent tidewater epidemics. He agreed with Rush on the need to plant crops and with Currie on the need to keep the marshes flooded. Basing his opinion on the work of Lavoisier and Ingenhousz, he explained how aquatic plants made the air healthy. At this point he split with his learned colleagues and concluded: "Heretofore mankind seem to have viewed their [marshes'] existence as noxious to them and unnecessary to their happiness. . . . I consider them as very necessary to keep the atmosphere in a proper degree of purity, for it is not only the impure atmosphere which kills animals, but the too pure also." Marshes, Seybert exhorted his fellows, "appear to me to have been instituted by the Author of Nature in order to operate against the powers which vegetables and other causes possess of purifying the atmosphere."

[34] Malthus, "A Summary View of the Principle of Population" in *Introduction to Malthus*, ed. D. V. Glass, pp. 159–68.

[35] Thomas Wright, "On the Mode Most Easily and Effectually Practicable of Drying up the Marshes of the Maritime Parts of North America," *Transactions of the American Philosophical Society* 4 (1799): 243–46; William Currie, "An Inquiry into the Causes of the Insalubrity of Flat and Marshy Situations; and Directions for Preventing or Correcting the Effects Thereof," *Transactions of the American Philosophical Society* 4 (1799): 139, 141, 142.

With an eccentric and prophetic remark concerning the role of marshes in the "economy of nature," this forgotten physician and statesman surmised, "Ere long marshes will be looked upon by mankind as gifts from heaven to prolong the life and happiness of the greatest portion of the animal kingdom." [36]

In the face of centuries of human toil, advancing technology, and growing mastery of insular seas, Seybert suggested that perhaps marshes should remain uninhabited in order to correct the "too pure atmosphere." While his reasoning was faulty, Adam Seybert early introduced the idea of the physio-chemical necessity of marshes in the preservation of coastal wildlife. Yet the probability that human modification of marshes might endanger estuarine wildlife did not become a widespread concern because the typhoid and yellow fever epidemics of the crowded coastal towns were thought to originate in the marshland's miasma or bad air.

Both urban commercial development and agricultural expansion were based on ideals that were counter to Seybert's assumptions concerning the necessary role of marshes in nature's economy. The ideas of Jared Eliot, an early advocate of soil conservation, demonstrate the dominant attitudes of both urban and rural eighteenth-century America. "A Swamp in its original Estate," he declared, is "full of Bogs, overgrown with . . . Brakes, poisonous Weeds and Vines . . . the genuine Offspring of stagnant waters." Eliot described the plant life of wetlands as "baleful Thickets of Brambles" and the animal life as "creeping Verm'n." Unlike Seybert, Eliot suggested that marshes were "the Dwelling-Place . . . of every unclean and hateful bird." An advocate of soil improvement, Eliot praised the drainage of such areas for agriculture as "a wonderful Change" and "pleasing to the Eye." As a representative of the prevailing eighteenth-century American attitudes toward wetlands, Eliot considered the clearing, drainage, and cultivating of swamps as "the happy Product of Skill and Industry." [37] Such ideas when translated into action left an agrarian imprint on reclaimed tidelands. Urban structures and rural farmlands alike increasingly occupied publicly entrusted tidelands.

Even prior to the nineteenth century the necessity of marshes was disregarded in favor of a continued reliance on estuarine dredging and coastal reclamation. As the image of the garden symbolized the eighteenth-

---

[36] Adam Seybert, "Experiments and Observations on the Atmosphere of Marshes," *Transactions of the American Philosophical Society* 4 (1799): 415–28.

[37] Glacken, *Traces on the Rhodian Shore*, pp. 692–93.

century intellectual recognition of humanity's role in taming and cultivating wild landscapes,[38] the creation of a seaside agrarian environment encouraged the replacement of swamps and marshes with diked meadows or farmlands. Urban growth, too, fostered dumping, drainage canals, wharves, and commercial shoreline development. As diseases repeatedly swept the crammed and dingy tidewater cities, the pastoral image of the garden became a paramount influence on scientists and writers, relegating noxious and obstructing coastal wetlands to a role as inefficient backwaters.

[38] Leo Marx, *The Machine in the Garden*, p. 73.

3

The Legacy of the Seashore Naturalists

There is at all times and in all places a fascination in the seashore,
with which we explore the rocky precipices of Mount Desert or
follow the sandy cliffs of Long Island.

—Oliver Bunce, 1872

WIDESPREAD changes in nineteenth-century landscapes wrought by canals,
manufactures, and agriculture encouraged the view of marshes as an un-
healthy encumbrance to the progress of society. Numerous alterations
brought about by new inventions encouraged the popular notion that engineering improved geography or otherwise completed an "unfinished"
landscape.[1] Marshes could be reclaimed with no apparent injury to the so-
cietally valued parts of nature. In a country with an abundance of unim-
proved land and a dearth of labor, the task of domesticating the landscape
was very challenging. Such an opinion was held by a European visitor,
Francis Bailey, who later became a cofounder of the Royal Geographical
Society in 1830. While crossing the New Jersey meadowlands in 1796,
Bailey noted that "over this swamp they have made a causeway, which . . .
shows how far the genius and industry of man will triumph over natural
impediments."[2] Although an old device, the causeway was a symbol of the
new reliance of the nation on overland transportation, which supplanted
rivers and estuaries as the primary arteries of communication, commerce,
and travel. Commercial dependence on land transportation accelerated the
reclamation of tidelands in the urban east. Drainage and dredging costs
were vastly reduced by the use of steam engines in reclamation.

In addition to the widespread adaptability of the steam engine to reclamation and transportation, an important factor that shaped attitudes of

---

[1]Clarence Glacken, *Traces on the Rhodian Shore*, p. 192; John Passmore, *Man's Responsibility for Nature: Ecological Problems and Western Traditions*, pp. 32–40.

[2]Francis Baily, *Journal of a Tour in Unsettled Parts of North America in 1796 and 1797*,
ed. Jack D. L. Holmes, p. 32.

early nineteenth century Americans toward tideland drainage was the prevailing ethos of Jeffersonian democracy. This ideology valued agricultural
expansion and a limited role for the federal government in the affairs of the
nation. These ideas influenced the future of coastal wetlands in two ways.
Jeffersonian ideology both encouraged reclamation for agriculture and afforded the individual states an opportunity to control development of the
waterfront by making grants of tidelands to facilitate commerce. The
events in Boston are but one example of the widespread conversion of tidelands for wharfage and manufactures. An early conflict between the public's access to the shoreline and reclamation to further navigation can be
discerned in the incremental filling of Boston's peninsula. Jefferson's followers generally facilitated agricultural and commercial reclamation, but
they also initiated the earliest federal surveys of the coastal seas, employ-
ing naturalists and surveyors to make charts of estuarine harbors.

As the pace of reclamation increased along the northeastern waterfronts, a few individuals suggested that quiet stretches of seashores, including some tidal marshes, be protected as sanctuaries or parks. Although
there had been colonial precedents for local wildlife protection, nineteenth-
century sympathies reflected the body of art, literature, and nature studies
created by the romantic temper. Some artists and writers extolled the rare
beauty of seashores and depicted the tidelands and marshes as essential to
birdlife, while naturalists studied water quality and quantity as significant
influences on the vegetation and wildlife of estuaries.

Many Americans, however, saw positive returns from the reclamation
of tidal flats and marshes, because conversion of wetlands provided homes
in congested cities or new farmlands with easier access to urban markets.
The impressive size and extent of coastal wetlands along the Atlantic seaboard encouraged this dominant attitude. Although isolated examples of
conflicts between fishing and navigation in the tidelands arose, only the
naturalists appreciated the long-term incompatibility between these competing uses of coastal lands and waters. To the few naturalists familiar with
Atlantic estuarine wildlife, the decline or disappearance of oysters, shad,
or salmon fisheries were the earliest indicators of the adverse effects of
dams, dredging, drainage, and reclamation on the spawning habits of these
species. But the lasting contribution of those artists, naturalists, and poets
who turned their attention to estuarine shores was to associate marshlands
with feelings of expansive freedom—to regard them as places for solitude
and contemplation and to use them as a symbol of the eventual imperma-
nence of coastal reclamation.

The naturalist's most important bequest to the formulation of an estuarine preservation ideal was a catalogue of plant and animal species to provide rudimentary evidence in support of Adam Seybert's earlier speculation that marshes were essential to the health of all species. The slow popular acceptance of these protectionist sympathies is best understood in relation to the economic development and technological advances that transformed the northeastern tidewater. In addition to rapid urban growth and increasing manufacturing, Jefferson's influence hindered any sustained and coordinated policies for coastal protection. Jefferson and his followers shaped the country's early ideas concerning land distribution, federal aid to economically ailing fishermen, federal encouragement of a national system of roads and canals, and the federal sponsorship of a coastal survey. Each of these proposals encouraged public financial support for reclamation as one of several means to achieve personal and national economic security.

In the preliminary stages of economic development, roads and canals formed an interdependent grid that particularly influenced the uses of coastal wetlands. E. I. Du Pont summed up the effects of economic growth in the vicinity of Wilmington, Delaware, an estuarine port where roads benefited commerce. "New houses, new wharves are built in town every year; two fine bridges have been constructed," wrote Du Pont, who boasted, "The face of the country is changed." His letter, written to Isaac Briggs, a surveyor, inventor, teacher, and bureaucrat in the Jefferson administration, described five new turnpikes that opened the Philadelphia market to Wilmington's farm produce. This underscored the slow shift from reliance on waterways to overland travel.[3] Changes in transportation altered the role of estuaries from a nexus in colonial times to a terminus in modern times.

This gradual shift led to a new dependence on the older services provided by estuaries for waste disposal and navigation. By comparison, the reliance on marshes for thatch, hay, and food gathering became marginal. As technological change gathered momentum, new inventions facilitated the reclamation of tidal marshes and inland swamps. Steam engines were used to saw timber, and eventually early railroads traversed the swamps to log previously inaccessible cypress stands.

An important invention demonstrated by Oliver Evans in Philadelphia

---

[3] Letter from E. I. Du Pont to Isaac Briggs, December 30, 1815, quoted in *Readings in Technology and American Life*, ed. Carroll W. Pursell, Jr., pp. 36–38.

in 1805 aided both commercial expansion and the filling of tidal marshes. Evans was a mechanical genius far ahead of his era. His completely automated flour milling system and high-pressure steam engine advanced eighteenth-century agriculture and manufacturing. Since rivers were used as waste-disposal sites, garbage had obstructed the Water Street docks along the Schuylkill River in Philadelphia. Evans constructed a self-propelled, steam-driven carriage that functioned as a dredge when floating. His dredge utilized a chain of buckets run by a high-pressure steam engine to clear the debris from the wharves more efficiently than hand-operated models. Christened the *Orukter Amphibolos* by Evans, the flatboat had been commissioned by the City Board of Health as yet another response to the yellow fever epidemics of the previous decade.[4]

Steam dredges like the Evans prototype were repeatedly improved during the nineteenth century. These devices eased the recovery of oyster shell deposits used in the construction of early roads, as well as mud, sand, or gravel. Eventually the centrifugal pump made the economics of dredging and reclamation more efficient, greatly increasing the water-lifting capacity of the more widely used Archimedean screw or wind-driven scoop wheel. In 1818 the Massachusetts pump appeared as a prototype of the centrifugal pump. The centrifugal pump played an important role in reclamation after 1867, when Henri Bazin, a hydraulic engineer for the Suez Canal, demonstrated its use in dredging as well as pumping. This increased the efficiency of pumping water, sand, mud, or gravel out of canal and harbor channels, creating dredge spoils or new land on what had been tidal marshes and mud flats. By 1877 suction dredges powered by steam-driven centrifugal pumps were used to maintain navigation channels on the Mississippi River. Between the introduction of Evans's steam dredge and the steam-powered suction dredges, the siltation of Atlantic estuaries by upstream erosion was forestalled by more efficient technology. Such navigational aids were essential to maintaining commercial advantages that the Atlantic coast's numerous safe estuarine harbors had always provided. Erosion caused by settlement and dumping increased the deposition of sediments in estuaries, choking channels and usurping older wharves.[5]

The vigilant protection of navigation was a consequence of expanding

---

[4] Arlan K. Gilbert, ed., "Oliver Evans' Memoir 'On the Origin of Steam Boats and Steam Waggons,'" *Delaware History* 7 (September, 1956): 159–62.

[5] Norman Smith, *Man and Water: A History of Hydro-Technology*, p. 36; J. Allen, "Hydraulic Engineering," *History of Technology*, ed. Joseph Singer, V, 535, 540.

frontiers, growing urban populations, and the prevailing attitudes of political economy. In 1803 five million Americans clustered east of the Appalachian Highlands in what French geographer Count Volney had called the Atlantic Country. Within sixty years the population increased to over thirty-one million persons in the Mississippi Valley, the Gulf and Pacific coasts, and the Atlantic tidewater. While the density of the entire nation's population barely doubled in this period, the density of the North Atlantic region increased fourfold, and in New York State alone density increased seven times. Because this expansion was concentrated in certain urban centers located on estuaries, environmental changes caused by drainage and reclamation were most obvious along the northeastern seaboard.[6]

Urban growth was contrary to the hopes of Thomas Jefferson, who shaped national policies to encourage agricultural settlement. He subscribed to a set of beliefs derived from Locke, Montesquieu, and Montaigne acknowleding the dominant roles of the physical and social environments in shaping human institutions. Specifically, Jefferson felt that cities brought out society's worst traits, whereas the wholesome influences of the agrarian countryside bred virtue, strength, and self-reliance. Accordingly, Jefferson influenced the wording of the Land Ordinance of 1785 establishing a uniform method of land surveys that divided areas into townships and ranges. He hoped this system would encourage the spread of small family farms. The ordinance applied to the unoccupied areas of the newly donated public domain. The larger seaboard states had relinquished their extensive claims to western lands between 1780 and 1785 in order to gain the necessary support of smaller states for the Articles of Confederation. This action represented the political birth of the public domain, administered by the legislative branch of a new central government.[7]

In order to encourage settlement, the ordinance reserved certain sections in each township for the support of education and military needs, while the federal government retained salt springs, mines, and mill sites. In theory this protection of certain lands to promote the national interest resembled the colonial precedent of using tidelands as a commons to promote fishing, hunting, grazing, or navigation. However, in practice the

---

[6]Ralph H. Brown, "The Land and the Sea: Their Larger Traits," *Annals of the Association of American Geographers* 16 (September, 1951): 200.

[7]Merle Curti, *The Growth of American Thought*, pp. 121–23, 165–69; Stewart Udall, *The Quiet Crisis*, pp. 25–36; Roy Marvin Robbins, *Our Landed Heritage: The Public Domain, 1776–1936*, pp. 7–10, 423.

federal retention of national-interest lands, except in the case of educational allotments, was a failure. The prime concern of Congress became the sale of public lands in order to provide revenue for governmental expenditures.[8]

Tidelands, however, represented a complex problem of public administration depending on the coastal territory in question. The proprietary responsibility for these wetlands that had existed in the colonies was derived from the crown, to be held in perpetual trust for the promotion of navigation and the public privileges of fishing and hunting. These powers were relinquished to the individual colonies as a condition of their sovereignty—granted by the Treaty of Paris of 1783. Two conflicting interpretations were now possible. Following the precedent of colonial times, the tidelands of the public domain ought to be the trust of responsible state governments. With the adoption of the Northwest Ordinance of 1787 and the later adherence of the Constitution to that ordinance, control of the territories was vested in Congress until the time of statehood. The Constitution of 1789 further complicated this state trusteeship by granting Congress plenary power over interstate commerce or navigation. Thus, land lying under a navigable waterway was held by either states or private parties, subject always to the public right to facilitate commerce. Should any willful obstruction of this public right-of-way occur, the federal government was authorized to remove the offense. While this interpretation had not crystallized in early nineteenth-century jurisprudence, it remained a widespread belief of nationalists, who asserted federal authority over local concerns.[9]

Jefferson's influence on tidelands was also felt in his opposition to subsidies to fishing interests and his uneven encouragement of the coast survey and support for roads and canals. Despite his interest in science, as secretary of state, Jefferson preferred an economic solution to what he himself admitted was a deplorable situation. He commented in 1791 that

[8]Garrett Hardin, "The Tragedy of the Commons," *Science*, December 13, 1968, pp. 1243–48; Benjamin Horace Hibbard, *A History of Public Land Policies*, pp. 2–58; Robbins, *Our Landed Heritage*, p. 8; Joseph M. Petulla, *American Environmental History*, pp. 76–77; Malcolm J. Rohrbough, *Land Office Business: The Settlement and Administration of American Public Lands, 1789–1837*, pp. 13–15.

[9]*Martin* v. *Waddell*, cited in U.S. War Department, United States Army Corps of Engineers, *Shore Control and Port Administration*, pp. 27–28. See also *Gibbons* v. *Ogden*, 9 *Wheat* 1 (1824); Curti, *Growth of American Thought*, p. 23; Rush Welter, *The Mind of America, 1820–1860*, pp. 88–89, 116–17, 125, 162–63, 374–75, 389–90.

"our fisheries, . . . in spite of their natural advantages, give us just cause for anxiety." Jefferson viewed the decline in fish catches as a problem of foreign competition, and he blamed the consequent depression in commercial fisheries on the loss of British markets. His remedy was to negotiate new trade agreements rather than to provide federal aid to either the fisheries or unemployed fishermen, even though he recognized that among the Atlantic nations only the United States did not provide such governmental assistance to fisheries. His attitude toward combining military, commercial, and scientific motives in sponsoring the Lewis and Clark expedition to the Columbia River showed no such reticence.[10]

This expedition provided a precedent for Jefferson to further the same motives in his recommendation of "a survey of the coast" in 1807. Natural science surveys, he realized, formed the basis of the military and commercial powers that sustained the federal union. The history of the coast survey is pertinent to coastal tidelands and subsequent harbor improvements. The jurisdiction of the survey as provided by the act of February 10, 1807, included "the islands and shoals with roads or places of anchorage, within twenty leagues of any part of the coasts of the United States." The president placed it under the authority of his secretary of the treasury, Albert Gallatin, who hired the services of Swiss emigré Rudolph Hassler. Hassler soon found it was impossible to undertake the celestial observations for a comprehensive charting of the Atlantic shores. Not until 1811 was Hassler dispatched to Europe for the instruments, and with the intervening war he did not return until 1815. The following year actual surveying began with New York Bay and then spread out northward and southward. Congress, though, ceased to supply the money to run the survey, halting all work on April 14, 1818. The charts, instruments, and other information concerning the survey were turned over to the navy. No systematic and comprehensive survey was conducted by a civilian agency of the central government until the coast survey's revival in 1832. This resulted from stiffening resistance to the Jeffersonians' national plans from champions of a limited federal

[10]Thomas Jefferson, "Report on Fisheries and Their Decline and Foreign Competition," 1st Cong., 2d sess., February 1, 1791, in *American State Papers: Documents, Legislative and Executive of the Congress of the United States*, VII, 8–12; Reuben G. Thwaites, ed., *Original Journals of the Lewis and Clark Expedition, 1804–1806*, 8 vols. (1904. Reprint, New York: Arno Press); Donald Jackson, ed., *Letters of the Lewis and Clark Expedition with Related Documents* (Urbana: University of Illinois Press, 1962).

government or states' rights, as embodied in the Virginia and Kentucky resolutions of 1798 and 1799.[11]

Albert Gallatin preferred an active central government to encourage a truly national system of transportation. In his 1808 "Report on Roads and Canals," Gallatin attributed the difficulty of their construction to a scarcity of capital and labor in relation to the geographical distances involved. His plans for a comprehensive network of roads and waterways recognized the economic advantages of the East Coast's natural features. The plan stressed the necessity of safe inland shipping from Boston to Georgia and suggested that an intracoastal waterway behind the barrier islands of the outer coast could be linked by four "great canals along the Atlantic seacoast." To link these estuaries with the necessary canals, Gallatin recommended "the early and efficient aid of the federal government." Broadly conceived by Gallatin, this system of roads and canals would "unite, by a still more intimate community of interests, the most remote quarters of the United States." As a nationalist, Gallatin argued that the Jeffersonians ought to develop a national transportation system utilizing hydraulic engineering techniques to overcome natural obstacles to navigation. Gallatin argued that "no other single operation within the power of Government can more effectively tend to strengthen and perpetuate the union. . . ," and he lamented that the construction of the Delaware and Chesapeake canal was "now suspended for want of funds." The eventual successes of privately funded canal ventures near Boston and Charleston ultimately led to canal building that included the Erie Canal in 1825 and the later connection of the New York, Delaware, and Chesapeake bays.[12]

These projects demonstrated the commercial advantages inherent in navigational improvements. Growing urban markets created a demand for the conversion of tidal basins and marshland grazing areas into diked farmlands, as well as the reclamation of tidelands for manufacturing, residential, and commercial uses. Boston, having tripled in population from 1789 to 1824, provides an example of the pressures for profit that were

[11]Gustavus A. Weber, *The Coast and Geodetic Survey: Its History, Activities and Organization*, pp. 2–7.

[12]Albert Gallatin, "Report on Roads and Canals," Misc. Doc. No. 250, 10th Cong., 2d sess., April 4, 1808, *American State Papers* (1832), VII, 723–41; Hans J. Huth, *Nature and the American: Three Centuries of Changing Attitudes*, pp. 73, 81–82, 105–106, 113–19; Ralph H. Brown, *Historical Geography of the United States*, pp. 103–97; Petulla, *American Environmental History*, 114–18.

characteristic of the period. Early in the colonial era, settlers had utilized the town cove for commercial wharfage and the mill pond as a tidal-mill collecting basin, diked off from the Charles River. Sluices allowed the flood tide to be held behind the dam so that at the ebb water could be used to operate the mills.

The heirs of John Hancock agreed to sell a portion of Beacon Hill to the company chartered to fill in the mill pond in 1804. As a consequence of the destruction of the Mill Pond, new plans were put forward for the creation of a tidal basin and mills on the mud flats that stretched between Roxbury and the Boston peninsula—otherwise known as Back Bay. The first in a series of proposals was submitted to the legislature in 1813.

The plans were not without vitriolic critics. One, in a letter to the editor, scolded his neighbors: "Citizens of Boston! . . . have you ever inhaled the Western breeze, fragrant with perfume, refreshing every sense and invigorating every nerve? What think you of converting the beautiful sheet of water which skirts the Common into an empty mud basin, reeking with filth, abhorrent to the smell, and disgusting to the eye." This early defender of wetlands concluded, "By every god of sea, lake, or fountain, it is incredible." With less than 10 percent of the Massachusetts legislature voting, a bill was passed to incorporate the Boston and Roxbury Mill Corporation for the purpose of damming the 660 acres of Back Bay to create a tidal basin for powering new mills. Opened in 1818, the dam, a mile and a half long, carried a fifty-foot-wide toll road, which provided Boston with another link to the south shore of the Charles River. When the water had drained from the basin at ebb tide, the flats dried up, blowing fine dust everywhere. The nuisance was so great that additional sluiceways were constructed to keep the mud flats flooded at all times.[13]

The coming of the railroads to Boston from the south and west created such a significant alteration of Back Bay that the needed causeways led to the bay's total filling. The crisscrossing of the old tidal basin seriously threatened the already languishing milling industry by blocking the flow of water. Similar changes occurred in South Cove, where seventy-five acres of mud flats were covered with tracks, wharves, and warehouses. The remaining hills of the peninsula were trundled off as fill for either South Cove or the Charles River, so that only Fort Hill remained in 1861. By 1844 the grandest engineering plan for converting the Back Bay flats into a resi-

[13] Walter Muir Whitehill, *Boston, A Topographical History*, pp. 73–94.

dential district had been laid, and they led to filling the entire area by the late seventies. In 1849 the Massachusetts Board of Public Health entered the list of advocates of filling, describing the situation as "one of nuisance, offensive and injurious to the large and increasing population residing upon" Back Bay. Winslow Homer captured one consequence of this reclamation in a lithograph he did for *Ballou's Magazine* in 1859. As the carts of rubbish were dumped, scavengers from the surrounding towns picked their way through the debris to recover whatever useful items had been carelessly tossed away.[14]

Another use of steam besides in reclamation was routine steamer service to coastal and river towns. Once too distant from urban centers, outlying coastal areas became accessible relaxation grounds for those fortunate enough to have fled the confusion and swelter of the tidewater cities, especially during the humid summer months. The beaches of Nahant, oceanward of Boston and Long Branch on the South Jersey shore, had attracted visitors since the 1790s. Cape May, New Jersey, and Long Island's Rockaway Beaches were so frequented by Philadelphians and New Yorkers, respectively, that boarding houses were built before 1812. Regular summer steamer service between Philadelphia and Cape May began in 1819, with a second line shortly thereafter. One result of this burgeoning of seaside resorts was the discontinuation of certain South Jersey farm holidays. One held on the third Sunday in August was called Big Sea Day, and the farmers of Seagirt used to hold an additional festivity in August called Little Sea Day.[15]

By the Jacksonian era, certain values became associated with outings to the seashore, including freedom, the restoration of health, and the appreciation of natural beauty. Although many of Boston's young lovers frequented the mill dam and other bridges to escape the ever prudish gazes of their elders, seashore excursions became particularly associated with a certain permissiveness. One French visitor to Nahant in the 1820s was surprised to see scantily clad women swimming and playing billiards. Both of these pastimes were considered proper only for men. During the 1830s and

---

[14] Ibid., pp. 141–73.

[15] Huth, *Nature and the American*, pp. 112–15, 118; Richard V. Anderson, "The Cape May Boats," *Cape May County Magazine of History and Genealogy*, 1955–63, pp. 55–62; Ellis Voss, "Summer Resort: An Ecological Analysis of a Satellite Community" (Ph.D. diss., University of Pennsylvania, 1941), pp. 18–19; Maria Leach, ed., *Standard Dictionary of Folklore and Mythology* (New York: Funk and Wagnalls, 1949–72), p. 140.

1840s, summer homes were built at Nahant in addition to boardinghouses. Theologian Timothy Dwight explained the attraction that these promontories and beaches held for many visitors and residents. "The monitory voice in the sea," he suggested, "affects the old and even accustomed mind with awe" and "makes an impression no human power . . . can ever entirely deface." [16]

For the remainder of the century visitors to Nahant echoed Dwight's sentiments. One contributor to William Cullen Bryant's edition of *Picturesque America* in 1872 remarked that "all the loveliness and majesty of the ocean are displayed around the jagged and savage browed cliffs of Nahant." As the marshy shores of the Boston peninsula were usurped for commercial and residential space, the city's population sought solace in the more imposing shoreline of Nahant and the Eastern Shore from Cape Cod to Portland, Maine. This same writer declared that "there is no coast on the Atlantic seaboard which presents a wider choice for the lovers of marine pleasures." [17] Those pursuits included recreation, art, camping, health, and bathing. By the end of the century, bicycling had become a popular way to tour the coast. Edmund H. Garret advised his readers to bicycle this coast because "it is the very closeness of the Nahant Cliffs to . . . cities" that "freshens and magnifies the impression produced by the ocean." Written in 1897, after a century of expansion, urbanization, and technological advance, Garret's book reflected a psychic need for picturesque landscapes. Of the less spectacular tidelands around Lynn, he wrote, "The slopes and marshy places are splashed broadly with golden rod and tansy, with the rich red of Joy Pye weed and the somber purple of ripe elderberries." [18]

These sentiments concerning the Massachusetts coast were part of a trend in the nineteenth century that was heavily influenced by the romantic movement in art, literature, music, and philosophy. While the diverse sentiments reflected in the romantic mood were inconsistent and defy facile explanations, the romantic appreciation for nature included certain types of coastal preservation. Parks along beaches or rugged promontories were most popular. Protection of marshlands was not a widespread or significant feature of romantic thought, but a few artists, poets, and naturalists turned their attention to coastal wetlands and left a positive view of marshes from

[16]Huth, *Nature and the American*, p. 113.

[17]William Cullen Bryant, ed., *Picturesque America: Or the Land We Live In*, II, 395, 399.

[18]Edmund H. Garret, *Romance and Reality of the Puritan Coast*, pp. 50, 64.

which later generations drew inspiration for preserving estuarine environs. The romantic tradition's preoccupation with unspoiled nature, its apprecia- tion of the sublime, its delight in picturesque settings, and its passionate sense of wonder challenged the customary dismissal of tidal marshes as wastelands. During the period when most Americans became appreciative of the outer coast, a few enthusiasts suggested that estuarine wetlands also afforded a constant reminder of the ocean's awesome powers.[19]

Although the roots of romanticism lay in Europe, an early American exponent of this tradition was naturalist William Bartram. He was the son of the "King's Botanist" John Bartram, whose exotic and native plant col- lections formed the colonies' first botanical gardens, located on the Schuyl- kill River in Philadelphia. For well over a century, the travels of John and William Bartram to the southeast were emulated by countless writers, art- ists, and military officials. William did more than describe exotic plants and animals. A sense of wonder and awe for wild landscapes filled his catalogue of new species.

As heir to a botanical reputation, William Bartram was surrounded by his father's natural history circle. This included the colonial naturalists Cadwallader Colden of New York and Alexander Garden of Charleston and their important sponsors in Great Britain. William eventually influenced romantic writers in England and America and furthered the course of na- ture study in the young republic. The family's friends included Benjamin Franklin, Thomas Jefferson, and Charles Wilson Peale. William later aided the careers of Alexander Wilson, a founder of American ornithology, bota- nist Thomas Nuttall, and entomologist Thomas Say.[20]

William's understanding of coastal wetlands began on a journey with his father to the Carolinas and Florida in 1765. William spent from March 20, 1773, until New Year's Day, 1777, touring the coastal and interior areas of the southeast. In 1791 he published a critically acclaimed account of these travels, inspiring such romantic writers as William Wordsworth and Samuel Taylor Coleridge. Bartram's reports of fish and bird abundance bear a striking similarity to the journals of earlier explorers, but in his work the earlier traces of an American appreciation for nature reached a deeper

---

[19] Roderick Nash, ed., *The American Environment*, pp. 44–67; Huth, *Nature and the American*, pp. 34–53, 148–77; Andrew Jackson Downing, *The Architecture of Country Houses* (Philadelphia: G. S. Appleton, 1850), p. xx; Charles H. Hosmer, *The Presence of the Past: A History of the Preservation Movement*, pp. 40–62, 76–133.

[20] Joseph Kastner, *A Species of Eternity*, pp. 3–4, 40–67, 79–113.

maturity. Primarily in his descriptive passages, Bartram introduced the world to the beauty of Georgia's Okefenokee Swamp and the Sea Islands, in addition to Florida's wetlands.[21]

Bartram reflected the influence of the English botanist John Ray, who had published *The Wisdom of God Manifested in the Works of Creation* in 1691. Ray offered a revived view of the natural world as a repository of God's virtuous intent and a reflection of the Creator's morally ordered mind.[22] Like his contemporary, Adam Seybert, William Bartram was convinced that wild areas were sanctuaries for necessary or useful species. Coastal wetlands were to these investigators essential areas for collection or experimentation in order to unravel the meaning of creation.

In spite of the preponderant American attitudes viewing tidal and river marshes as wastes of sand, mud, and slime, William Bartram conceived of these areas as natural gardens wherein flourishing wildlife displayed the beauty, diversity, and divinity of all creation. While many writers advocated human struggles to control nature as a desire to replace savagery with quiet pastures or orderly villages, Bartram found a certain solace in the wild southeastern swamps and coastal wetlands.[23]

Further, Bartram associated the timeless scenic qualities of the seashore with unmatched evocative powers offering humans a chance to transcend temporal existence. For example, in describing his entrance to Charleston by ship, Bartram voiced a refrain important in the works of later romantic writers. "As we approach the coast, . . . the amplitude and magnificence of these scenes are great indeed, and may present to the imagination an idea of the first appearance of the earth to man at creation." [24]

To the elites of the eastern seaboard, the works of Bartram and later romantic artists and writers offered a counterview to ideas that estuaries were merely safe harbors and that the coastal wetlands were unfinished landscapes in need of reclamation. To a handful of educated townsmen, such landscapes were no longer the raw material for shaping a new society but were a sustainer of natural order and a reminder of God's wisdom.

[21] William Bartram, *The Travels of William Bartram*, ed. Francis Harper, p. 44; Huth, *Nature and the American*, pp. 14, 21.

[22] Kastner, *Species of Eternity*, pp. 82, 85.

[23] Robert Hanie, *Guale, The Golden Coast of Georgia*, p. 94; Kastner, *Species of Eternity*, pp. 94–95; Bartram, *Travels*, p. 78.

[24] Bartram, *Travels*, p. 2.

These contrasting sentiments created an ambivalence in the attitudes of sensitive observers concerning coastal wetlands. The tidelands of estuaries were sources of exploitable wealth, obstructions to vanquish through reclamation, or convenient places to contemplate the sea and study nature. This ambivalence was nurtured because the ocean retained its frightening and sublime qualities despite efforts to harvest its fish, chart its varied shores, or reclaim its extensive tidelands. Estuaries as landward intrusions of the ocean were a reminder of the terror of the primeval seas even as the cities of the Atlantic seaboard grew into sizable harbors.

Artist John Singleton Copley captured this dichotomous mood of the harbor in his painting *Watson and the Shark* in 1778. Having painted an actual occurrence, Copley depicted the sea naturalistically, as terrified fishermen struggled to fetch their comrade from the menacing jaws of an advancing shark amidst the background of a quiet harbor. Here the terror of the deep was contrasted with the safety of the harbor and graphically demonstrated the ambivalence of human views of the sea as both a haven and a graveyard. American attitudes concerning the sea were transferred to estuaries and their wetland fringe. Safe anchorage required constant navigational improvements in estuaries. Equally important was the filling of "pestiferous" estuarine wetlands for agriculture, commerce, and even seaside parks. During this first period of environmental concern for seashores, when many Americans sought out beaches for recreation and picturesque promontories for contemplation, national attitudes still compared wetlands with more majestic landscapes and found them wanting. However, a few artists, naturalists, and poets discovered a rare beauty in the wilderness of tidal marshes.

The diverse and sundry moods of the shore were portrayed by two New England artists, Washington Allston, who introduced the romantic style of American painting, and Winslow Homer, who popularized the seashore after the Civil War. Although Allston was born in the South in 1779, he was raised in New England and toured Europe in his youth. His painting entitled *Rising of a Thunderstorm at Sea* (1804) dramatically portrayed the ocean as sublime. In contrast, Allston explored the effects of light on the seashore in *Coast Scene of the Mediterranean* (1811). The careers of Allston and Homer helped attract numerous artists to the seashore as either a source of inspiration or as subject matter for their painting.

In the same romantic genre, Thomas Cole, an organizer of the Hudson River School of painting, visited Mount Desert Island, Maine, to bask in the sublimity of this awesome rockbound shore. Cole and his companion believed the island's eastern beaches were of unsurpassed beauty. As part of the larger national rediscovery of the picturesque landscape, the painting of marshes reached its fullest expression in Martin Johnson Heade's series of luminous studies beside Newport's marshes. Heade, like Allston, adopted a storm theme in a painting entitled *Approaching Storm , Beach near Newport* (1860).[25]

Winslow Homer had ample artistic company when he fled the crass materialism of the Gilded Age to live along the rugged coast of Maine and paint the many moods of the Atlantic shores. The popularity of Homer's works revealed a certain late-nineteenth-century American weariness with the pace of life and the degree of urban change. Close to the sprawling urban centers from Baltimore to Boston were numerous quiet seashore retreats that offered inspiration, relaxation, or physical exertion, depending on the needs of the visitors. Long Branch, New Jersey, was captured in Homer's paintings, and the lure of Coney Island or the Rockaways was reflected in the words of Walt Whitman. Many citizens shared Ralph Waldo Emerson's feelings when he said that "in the presence of nature a wild delight runs through the man, in spite of real sorrows." The proximity of the seashores to large urban populations in the northeast encouraged coastal reconnoitering of nature's picturesque and majestic aspects. Herman Melville asserted that, in the midst of the crowded streets of Manhattan, Battery Park at the tip of the island overlooking upper New York Bay afforded countless workers a daily respite. For Melville, water and meditation were wedded.[26]

The shore's allure was best expressed by the English poet George Gordon, Lord Byron, who wrote, "There is a rapture on the lonely shore . . . by the deep sea, and a music in its roar." Byron addressed one of the characteristics of the seashore that appealed to a romantic imagination. This was the widespread belief that human attempts to tame the earth through cultivation or reclamation were futile given the awesome power of the sea.

[25] Richard McLanathan, *Art in America: A Brief History*, pp. 116–23; Huth, *Nature and the American*, pp. 82–91, 118.

[26] Ralph Waldo Emerson, "Nature" (1836), in *Emerson's Works* (Boston: Houghton Mifflin, 1883), I, 13–14; Herman Melville, *Moby Dick: Or the White Whale* (New York: Airmont Publishers, 1964), p. 26.

Roll on, thou deep and dark blue ocean—roll!
Ten thousand fleets sweep over thee in vain;
Man marks the earth with ruin—his control
Stops with the shore.

Reiterating Bartram's description, he noted of the sea, "such as the creation's dawn beheld, thou rollest now," and then Byron fashioned a lasting
impression of the oceans: "Dark-heaving—boundless, endless, and sublime—the image of Eternity." [27]

Yet for all this apparent preoccupation with the sea's picturesque
beauty, the reclamation of tidelands continued. In fact, a major irony of the
romantic period was this glorification of nature while land and raw materials were rapidly, wastefully, and remuneratively used by widely respected
and eulogized entrepreneurs. The fur trade, for example, had destroyed
numerous seals, sea otters, and beavers when Washington Irving—whose
patron was John J. Astor—offered a romanticized ideal of the trapper in
The Adventures of Captain Bonneville and *Astoria*. A critic of industrialization, Irving depicted the hunter as possessing uncommon virtues in his
daily physical encounters on the frontier. [28]

Part of Astor's great wealth was derived from his grant of New York
state tidelands adjacent to his upland holdings. The grant was made in 1810
on the condition that Astor fill and bulkhead the submerged land for the
purposes of maintaining a wharf and warehouses. [29] This was a notable example of the effects of urban expansion, technological development, and
burgeoning population along the tidewater. These changes altered the
water quality, size, shape, and biotic relationships of the marshes, mud
flats, and oyster reefs in estuaries. Since state and local governments relied
primarily on property taxes for revenue, the government had additional incentive to develop tidelands through engineering. However, the reaction of

[27] Lord Byron, *Childe Harold's Pilgrimage*, Canto IV, stanzas 178–83 (London:
J. Murray, 1819).
[28] Suzanne Fries Liebetrau, "Trailblazers in Ecology: The American Ecological Consciousness, 1850–1864" (Ph.D. diss., University of Michigan, 1973), pp. 1–19, 202–43;
Hosmer, *Presence of the Past*, p. 40; Huth, *Nature and the American*, pp. 165–77; Washington Irving, *Astoria: Or Anecdotes of an Enterprise Beyond the Rocky Mountains*, ed. Edge-
ley W. Todd (Norman: University of Oklahoma Press, 1964), pp. xv–xliii; Vernon Louis
Parrington, *Main Currents in American Thought*, vol. 2, *The Romantic Revolution*, pp.
197–207.
[29] U.S. Army Corps of Engineers, *Shore Control and Port Administration*, pp. 46–47.

the knowledgeable elite to the construction of upriver dams, the drainage of marshes, and the siltation of harbors engendered a cultural reappraisal of technology and conservation in national life. Subsequent anxiety over industrialization and landscape alteration produced wildlife conservation, urban parks or recreation areas, public health movements, and scenic preservation.[30]

These views, however, ran counter to the materialist values of the developing nation. Essential to the developing American ethos were the protection of private property, the maintenance of limited government, and the rights and obligations of contracts. For the majority, any attempt to preserve marshlands from reclamation would have been viewed as a tyrannical usurpation of the right of individual owners to care for their property as they saw fit.[31]

As American scientific ideals were formulated, this resistance to the technological conversion of marshes into productive land would have been unacceptable. Yet the protection of locally endangered species was not beyond the interest of state legislatures. In 1709 Massachusetts forbade the construction of any obstacles to fish migration in its rivers. To counter enforcement difficulties all colonies faced, many authorized local powers to curtail poaching. Connecticut authorized seacoast towns to regulate the taking of oysters and clams in 1784. By the time the New York legislature protected the Long Island heath hen and other game in early 1791, the precedents were nearly a century old.[32]

The noted natural historian Dr. Samuel Latham Mitchell was a supporter of game, bird, and fish protection, and he lamented that "the life of the herring has never been thought important enough to put under the protection of the law in New York." A tireless naturalist and congressman,

---

[30] Within the "tidewater" region (Brown, *Historical Geography*, pp. 173–91) of the Atlantic coast, historic preservation of Mount Vernon and Christ "Old North" Church, Boston, as well as a scenic cemetery at Mount Auburn, was undertaken even before the Civil War. Tidewater Virginia and Connecticut contributed to agricultural conservation. However, seaboard America had not chosen to preserve tidal marshes; rather, their reclamation for housing and commercial purposes was considered civic improvement. The military had had virtually the only previous national experience with sequestering parts of the shore, and that for the construction of forts (Hosmer, *Presence of the Past*, pp. 41–122).

[31] Welter, *Mind of America*, pp. 107, 112, 139–40, 212, 237.

[32] Richard Hofstadter, *The American Political Tradition*, p. viii; Glacken, *Traces on the Rhodian Shore*, pp. 471–84; Benjamin Farrington, *Francis Bacon: Philosopher of Industrial Science*; Brooke Hindle, *The Pursuit of Science in Revolutionary America*, pp. 190–91; Merle Curti, *American Paradox: The Conflict of Thought and Action*.

Mitchell published the first ichthyological tract in the nation. *The Fishes of New York* (1815) contained descriptions of the life cycles of 116 species. Fish conservation so interested Mitchell that, while in the state assembly, he spoke at great length for the prohibition of a dam on the upper Hudson that would have blocked the migration of trout and Atlantic salmon, and his bill passed. Though an advocate of the Erie Canal, Mitchell represented the early attempts to balance technology and protection in the eighteenth century.[33]

As an avid Jeffersonian and unsurpassed savant of New York's land and sea environments, Mitchell, who served in Congress from 1801 to 1813, was called by Jefferson the "congressional dictionary." Besides being active in politics, Mitchell founded the New York Lyceum in 1817, encouraged his friend Dr. David Hosack in the establishment of Elgin Gardens, and taught chemistry at Columbia College. Commencing with his sponsorship of protection for the Long Island heath hen, Mitchell concerned himself with both bird and fish conservation throughout his remarkable career.[34]

Mitchell's work influenced amateur studies of geology, botany, and zoology in the early republic. The general approach of investigators who collected specimens for scientific study was called natural history, and consequently these practitioners were known as naturalists. Their careful attention to the details of fish and bird habits, water quality, and vegetational changes shaped the formulation of an estuarine preservation ideal. When their information was publicized by artists and writers, a slow understanding of nature and the need to protect endangered species spread among some of the population. Romantic writers provided an intellectual respectability for primitive life, untamed beauty, and historic values inherent in landscape. Thus, with regard to the seashore, sentiments for the preservation of monumental scenery, picturesque recreation sites, and wildlife refuges among the marshes rested on a concatenation of scientific, artistic, and literary contributions.[35]

The national expansion to the Gulf and Pacific coasts offered oppor-

[33]Ralph Merton Van Brocklin, "The Movement for the Conservation of Natural Resources in the United States Before 1901" (Ph.D. diss., University of Michigan, 1952), pp. 83–85.

[34]Kastner, *Species of Eternity*, pp. 193–206.

[35]Liebetrau, "Trailblazers in Ecology"; Roderick Nash, *Wilderness and the American Mind*, pp. 96–107; Donald Worster, ed., *American Environmentalism: The Formative Period, 1860–1915*, pp. 1–10.

tunities to naturalists to explore new areas and describe new species. For others, like John James Audubon, it encouraged a retracing of earlier men's travels. Audubon carried on the ornithological tradition of Alexander Wilson, an early popularizer of bird study. Jean Jacques Audobon was born in Haiti and raised in France before moving to Philadelphia. By 1806 young Audubon was working in New York for Dr. Mitchell, stuffing birds for the doctor's displays. This experience exposed Audubon to the growing natural history circle in New York City, which was soon to replace Philadelphia as the cultural and commercial center of the nation.[36] While Audubon did not then linger in New York, his later residence in upper Manhattan placed him in good company.

After an unsuccessful business venture in Kentucky, Audubon began his life's work, painting the birdlife of eastern North America and describing their habits. He began his work on October 12, 1820, and spent the next ten years searching for backers for his illustrated volumes of *The Birds of America*. He also encouraged his son and a close friend, Dr. John Bachman, of Charleston, to publish a work on native mammals for which the elder Audubon drew almost half of the plates. Audubon's travels included the old southeastern borders of Louisiana, Mississippi, and Alabama, the upper Missouri River Valley, and Labrador. His excursion through Florida from November 20, 1831, until May 31, 1832, produced many examples of estuarine and shore birds. Audubon's particular contributions to the formulation of an estuarine preservation ideal were to portray the birds in their habitats as national assets and to describe their feeding and nesting behavior.[37]

A major theme in Audubon's writings is the unimaginable abundance of common and exotic species along the Florida shores. Besides developing this theme, Audubon challenged William Bartram's characterization of Florida as a garden. For Audubon the landscape was dominated by "eternal labyrinths of waters and marshes." The waters of this frontier were "interlocked and apparently never ending; the whole surrounded by interminable swamps." Audubon represents a curious blend of nature enthusiast and sport hunter, displaying in his writings a personal ambivalence for wild nature. Describing his adventures "in a wild and desolate part of the

[36]Kastner, *Species of Eternity*, pp. 194–207.

[37]Catherine Hall Proby, *Audubon in Florida* (Coral Gables, Fla.: University of Miami Press, 1974), pp. 13, 102–103.

world," Audubon toiled against humid heat and insects to see birds in their most hidden locales.[38]

The differences between Audubon and Bartram reveal an ambiguity in the broader romantic persuasion concerning nature. Audubon reflected this contradiction by depicting the beauty and diversity of teeming birdlife against the backdrop of a fearsome wilderness. Audubon also represents a curious blend of nature enthusiast and sport hunter, displaying in his writings a personal ambivalence for nature. The attitude of the hunter was best expressed by Audubon in pursuit of his desire to paint the head of an alligator, for which it was necessary to obtain a specimen. "When within a few yards" of the animal, Audubon wrote, "one of us fired, and sent through his side an ounce ball, which tore open a hole large enough to receive a man's hand." In a haunting image of the reptile's expiration, Audubon recalled that the alligator "opened his huge jaws, swung his tail to and fro, rose on his legs, . . . and fell to the earth." Audubon also described a later encounter with an equally large alligator while he was sailing the river early one morning. "One of the officers," he remembered, "fired and scattered" the animal's "brain through the air"; then the beast "tumbled and rolled at a fearful rate, blowing all the while most furiously." These two incidents reveal the lingering appeal that wildlife destruction held for many Americans because such actions asserted human dominance in an otherwise inhospitable wilderness.[39]

Other aspects of Audubon's character made him an observant naturalist, whose attention to detail is reflected in his appreciation for salt marshes. Indeed, when they were lost on a trail not far from Saint Augustine, in the midst of a Florida rainstorm, Audubon's party had been led to safety by the odor of salt marsh air. Audubon excelled in the description of salt marshes and their birdlife. Florida's shores were the southern home of great marbled godwits, egrets, spoonbills, willets, great white herons, and great blue herons. The Louisiana heron was described by Audubon as "delicate in form, beautiful in plumage and graceful in its movements." This particular bird was highly prized for its plumage by the millinery trade, and within seventy years market hunters had made it nearly extinct. The herons' slaughter was an ironic consequence of nature appreciation in

[38] Ibid., pp. 87–91, 135.
[39] Ibid., pp. 191–200.

women's fashions. Audubon depicted the creature's allure, rendering the bird on a mud flat in front of a reeflike island covered with tropical vegetation.[40]

Audubon was among the earliest writers, and was perhaps the first artist, to introduce the tropical shoreline's vegetation to the national imagination. The mangroves used as the background or perch for the brown pelican and the Florida cormorant were described by Audubon as nature's reclaimers. The red mangrove, especially, he noted, "is very abundant along the coast of Florida and on almost all the Keys, except the Tortugas." Because of the shelter offered to creatures by the mangrove trees, varied forms of wildlife eked out a precarious existence on the thousands of islets forming the southern shore of Florida. Audubon accurately described the plants as "evergreen, and their tops afford a place of resort to various species of birds at all seasons, while their roots and submerged branches give shelter to numberless testacious mollusca and small fishes."[41]

Audubon carried on the naturalist tradition of Alexander Wilson and Samuel Latham Mitchell and their concern for birds and fish. His studies complemented the earlier appreciation for nature established by William Bartram. These investigators explained the relations among plants, birds, and fishes, leading to an important intellectual shift in biologically understanding estuaries and their wetland fringe. Especially through the art of Audubon, estuarine wildlife habitats became fixed in the literature, lore, and paintings of the romantic period. After three centuries of exploration and coastal reclamation, the birdlife of the tidewater began to be viewed not just as edible specimens but as beautiful symbols of nature's bounty.

Although the work of naturalists was generally still part of an ambivalent heritage concerning nature, a growing interest and precision in explaining shore biology was shown by naturalists Edmund Ruffin and Henry David Thoreau. The earliest systematic investigations of marsh habitats and soils and the effects of coastal landscape changes on vegetation were conducted by Edmund Ruffin on the outer banks of North Carolina from 1843 until 1859. As editor of the influential *Farmer's Register*, Ruffin was a respected advocate of pursuing soil conservation through an understanding of geology, chemistry, and botany. In 1832 he had published his thoughts on restoring soil fertility in *An Essay on Calcareous Manures*. His descrip-

[40] Ibid.
[41] Ibid., p. 220.

tion and examination of the seashore divided the barrier islands into five vegetational zones. Ruffin used high and low tide marks to distinguish the beach zone from the "sand flat," inundated by only the highest tides. These two areas comprise what is often termed the fore-shore. Inland from the sand flats were the dunes, which he called "high sand hills," covered with stunted shrubs and grasses. Behind the barrier dunes lay the live-oak and loblolly-pine forests bordered by the salt marshes near the bay.[42]

Ruffin presented these observations in a strikingly modern way. Ecologists would eventually call his classification of vegetation and its associated animals *zonation*. Discrete bands of life had adjusted differently to the stressful seashore environs. Wind and salt water regularly create stress for plants, and occasionally shifting sands can bury an entire cedar forest. Ruffin noted that the live-oak–loblolly-pine association was heavily browsed by animals, and the salt meadows were usually turned over to extensive stock grazing, as had been done since colonial times. His classification of organisms in relation to their surroundings was pioneering.

Furthermore, Edmund Ruffin recognized the effects of natural and man-made changes on the environment of the seashore. Ruffin described how plants adapt to their particular growing conditions through a process he called "acclimation." He recognized the tendency of loblolly pines to convert sandy soils with high moisture content into peat bogs because of their enormous fall of needles. Mosses grew among the pines' litter. For its ability to enrich the soil, Ruffin called the loblolly pine "one of the greatest blessings to our country."[43]

In addition to successional changes on land, he noted the deleterious effect of high-salinity water on North Carolina oyster reefs. Shifting sands and currents often allowed the sea to breach the barrier islands, and the resultant increase in sea water proved disadvantageous to brackish-water shellfish. These observations of water quality along the shore led Ruffin to speculate on the conditions necessary to promote malaria. While he did not go so far as to attribute it to fresh-water mosquitoes, he did note that the flooding of tidelands by fresh water in order to grow rice spread the malarial outbreaks.[44]

---

[42] Avery Odelle Craven, *Soil Exhaustion as a Factor in the Agricultural History of Virginia and Maryland, 1606–1860* (Urbana: University of Illinois Press, 1925), pp. 134–38, 139–40; Liebetrau, "Trailblazers in Ecology," p. 29.

[43] Liebetrau, "Trailblazers in Ecology," pp. 30–31.

[44] Ibid., p. 31.

Ruffin was also a critic of human "improvements" of the landscape. To him, the dams on streams for mill ponds and the flooding of tidal marshes for rice cultivation were as destructive to wildlife as the fires in pine forests. Although he did recognize the need for some levee construction and drainage, he suggested that individual laissez-faire behavior wreaked havoc on the community. Removal of the water from parts of one flood plain merely shifted it to a neighboring area, creating a new flood problem along the stream. In this particularly prophetic and environmentally conscious observance, Ruffin warned, "Should every proprietor exercise his equal right to embank all his own lands . . . the attempt must fail." The wisdom of that remark was noted by others and was graphically demonstrated by the settlement of Sacramento, California, in the 1850s and the subsequent thirty-year siege with the waters of its adjacent river. Similar floods occurred along the lower Mississippi Valley.[45]

Ruffin's feelings regarding human efforts to "tame" nature were revealed in his depiction of a dredge on the Chesapeake and Albermarle canal: "The seizing and tearing up of the roots and earth, and finally . . . the opening and emptying of its hand—all moved by the unseen power of steam—made the whole operation seem as if it was the manual labor of a thinking being, of colossal size, and of inconceivable power." While the admiration for technology was common to his generation, his recognition of the effects it had on landscape was not. "When our ancestors first reached this shore, nearly the whole country was in the state of nature," he explained in the manner of romantic writers of his day, who viewed Native-American cultures as passive in their relations with the environment. Referring to human "improvements," Ruffin said of early America, "No dams had obtructed the free and regular course of the streams and . . . no great and artificial floods were formed." His realization that human settlement provoked the very calamities that cultural evolution tried to mitigate places Ruffin among the early advocates of a land ethic.[46]

Ruffin's respect for unspoiled nature was shared by a New Englander of greater repute. Henry David Thoreau identified a priceless dimension of untrammeled nature: "We need this tonic of wilderness, to wade sometimes in the marshes where the bittern and the meadow hen lurk, and hear

[45] Ibid., p. 37.
[46] Ibid., pp. 40–41, 145, 153–54; Edmund Ruffin, *Essays and Notes on Agriculture* (Richmond: J. W. Randolf, 1855), pp. 265, 291, 331–32.

the booming of the snipe, to smell the whispering sedge where only some wilder and more solitary fowl builds her nest. . . ."[47]

Thoreau had been drawn to the seaside in 1849, when he began the first of several visits to Cape Cod. He referred to the shore as "a wild rank place . . . strewn with whatever the sea casts up, a vast morgue." Again and again he returned to the Cape in the decade before his death. There he wrote, "Everything told of the sea, even when we did not see its waste or hear its roar." The town of Orleans, he noted, "is famous for its shell-fish, especially clams," and he commented that "the shores are more fertile than the dry-land."[48]

As a sensitive observer, Thoreau did not overlook the historical changes wrought by civilization on the Cape's delicate ecology. He remarked in passing, "These were the 'plains of Nauset,' once covered with wood," but removal of the forest had forced the colonists to rely on peat for fuel. On another occasion, when visiting Wellfleet, he chanced to talk with oystermen and learned that "the native oysters are said to have died in 1770." Thoreau related the three commonly held theories accounting for their decline: ground frost, the pollution of the harbor by rotting blackfish carcasses, and—"the most common account"—that "Providence caused them to disappear." Thoreau lamented, "I find that a similar superstition with regard to the disappearance of fishes exists almost everywhere." Recalling the Native-American occupation of the area, he cited previous authorities who had attributed the original concentrated settlement of the Cape to "the abundance of fish." Having inspected several middens, Thoreau concluded, "The Indians lived about the edges of the swamps, then probably . . . ponds, for shelter and water."[49]

In countless passages Thoreau commented on the omnipresent force of the ocean. He wrote, "The seashore is a sort of neutral ground, a most advantageous point from which to contemplate the world." The ocean provided a certain reassurance concerning the relationship of humans to the sea. "Creeping along the endless beach amid the sunsquall and the foam," Thoreau mused, "it occurs to us . . . that we, too, are the product of sea-slime." To explain the inherent evocative power of the sea, Thoreau sug-

[47] Henry David Thoreau, *Walden* (New York: Airmont Press, 1965), pp. 222–23.
[48] Henry David Thoreau, *Cape Cod* (Boston: James R. Osgood & Co., 1871), pp. 28, 30–31.
[49] Ibid., pp. 51, 64, 74–77.

gested that seashore creatures were fascinating for their wilder qualities. "There is naked Nature—inhumanly sincere, wasting no thought on man—nibbling at the cliffy shore where gulls wheel amid the spray." [50]

Thoreau did not comment extensively on the Cape's numerous salt marshes, reserving his remarks instead for the enormous contrasts between the strange sublimity of the ocean and the scattered, familiar culture of the towns and villages. Yet his reference to the nature of life residing in the tidelands places Thoreau among the few naturalists who recognized the extraordinary character of intertidal life. He rightly associated shore creatures with those of the sea, where he believed that the animal and vegetable kingdoms "meet and are strongly mingled." Concerning the rigors of life in the tidelands Thoreau conjectured, "Before the land rose out of the ocean and became dry land, chaos reigned; and between high and low water mark . . . a sort of chaos reigns still." [51]

Thoreau's observations mixed a profound mystical reverence for nature with scientific accuracy. He noted the kinship between seashore and desert vegetation. He mentioned the affinity of ducks for the *Salicornia* (marsh samphire or pickleweed) flats and the use of the tidal marshes as common pasturage by the towns of Wellfleet, Orleans, and Eastham. Thoreau also identified the Cape as a faunal break for marine creatures, dividing the northern and southern shores of Massachusetts. Colder water species are replaced by warmer-tolerant clams, mussels, cockles, and periwinkles. Like Ruffin, he noted the vegetation zones along the shore and criticized the impact of civilization on the landscape. "No doubt there is some compensation for this loss," Thoreau noted concerning industrial progress, "but I do not at this moment see clearly what it is." During his trip on the Concord and Merrimack rivers in 1839 he criticized the mill dams along the Concord for their effects on shad migration. He grieved, "Poor Shad, where is their redress. . . ? Who hears the fishes when they cry?" Thoreau displayed the romantic appreciation for the cycle of renewal through birth, maturity, and decay. Such a philosophy comforted him when he observed the degrading effects of industry on the landscape. Concerning the Merrimack River, a center of textile and shoe manufacturing, he wrote, "Perhaps after a few thousand years . . . nature will have levelled

[50] Ibid., pp. 64, 94, 97.
[51] Ibid., p. 64.

the Billerica dam, and the Lowell factories, and the grass-ground river runs free again. . . ." [52]

In respect to the seashore and the origins of an estuarine preservation ideal, Thoreau was a striking blend of eighteenth-century and modern sensitivities. Like Seybert, he knew that marshes were necessary. Having read Wilson and Audubon, he was impressed by the affinity of certain animals to specific plants. He stressed the need to preserve the landscapes surrounding certain towns to enhance the character of a region. Despite the expense of landscape preservation, he felt it necessary, "for such things educate far more than any . . . system of school education." In recognizing the historical human impacts on landscape, Thoreau and Ruffin articulated very modern concerns for the capacity of nature to absorb the effects of reclamation. [53]

The interesting blend of philosophical idealism and descriptive nature studies inherent in romanticism engendered a positive reappraisal of marshes and estuaries. This change in attitudes also created a recognition of the adverse effects of dams, dredging, and reclamation on the wildlife of rivers and estuaries. The naturalists and later more specialized scientists uncovered support for Adam Seybert's intuitive grasp of the necessity of marshes. In so doing, these investigators formed a link between Seybert's theory and the findings of later ecologists. While artists and writers extolled the picturesque qualities of seashores and wildlife, the naturalists saw the seaside's equal potential for biological study.

Concurrently the public's attitude associated recreation and relaxation with an enjoyment of the seashore. Between 1790 and 1880 contradictory patterns characterized American response to the seashore. The conflicts inherent in later demands for coastal conservation can be traced to these early patterns. Because of the extent of coastal wetlands, during this period the demands for reclamation could be met in some areas while wildlife refuges remained in remote tidelands behind the barrier beaches.

Amidst efforts to reconcile competing uses of coastal areas, the growing sentiment for appreciating the wild recesses of the estuarine shores was best expressed in the poetry of Sydney Lanier. In 1878, he wrote the memorable poem "The Marshes of Glynn" about the landscape of Saint Simon's Island, one of the Georgia Sea Islands. Because he chose to write

[52] Liebetrau, "Trailblazers in Ecology," pp. 195–97.
[53] Ibid., pp. 196–97.

about a portion of the seashore that other romantics ignored or forsook for bolder coasts, Lanier made an important contribution to estuarine appreciation. Lanier accentuated the peaceful and contemplative quality of the tidal marshes. The poem described the marsh as a sanctuary for lovers or those wishing to temporarily ignore the day's chores, where the pace of life was slowed amidst the quiet and more serene ambience of the "marsh and the terminal sea." He poetically captured the dual nature of the tidal grasslands as "a world of marsh that borders a world of the sea." Lanier immortalized the evocative qualities inherent in these least appreciated national seashores. To him they resembled the prairies or great plains—"a league and a league of marsh-grass, waist-high, broad in the blade,/Green, and all of a height, and unflecked with a light or a shade." The *Spartina* or cord grasses dominating the tidal marshes of temperate shores transform the mud flats into a seaside prairie, within which Lanier found solace and inspiration. He viewed this land as "tolerant plains, that suffer the sea and the rains and the sun . . . how candid and simple . . . and free." [54] In the poem Lanier saw the tidal creeks and muddy paths as evidence of his Creator's handiwork: "Oh, like the greatness of God is the greatness within the range of the marshes, the liberal marshes of Glynn." He described the area as the sun was setting and the flood tide was approaching its peak. Lanier poeticized the successive zones of coastal vegetation that Ruffin and Thoreau had described more literally.

> Sinuous southward and sinuous northward the shimmering band
> Of the sand-beach fastens the fringe of the marsh to the folds of the land.

Lanier called the reader's attention to the dual nature of tidal marshes as the tide swelled:

> Look how the grace of the sea doth go
> About and about through the intricate channels that flow
>   Here and there,
>   Everywhere
> Till his waters have flooded the uttermost creeks and the low-lying lanes,
> And the marsh is meshed with a million veins, . . .
> . . . and all is still; and the currents cease to run;
> And the sea and the marsh are one. [55]

[54] Sidney Lanier, "The Marshes of Glynn" (1878), in *American Poetry and Prose: Part II, Since the Civil War*, ed. Norman Foerster (Boston: Houghton Mifflin, 1934), pp. 1059–61. For biographical information, see Foerster, p. 1054.

[55] Ibid., pp. 1060–61.

As in the paintings of Martin Johnson Heade, the tidal marshes were viewed as central to the sublime character of the seashore. They were not backdrops for more important scenes or allegories. Both Lanier and Heade used the play of light on these seaside savannahs to evoke a mood of silent reassurance and liberating expansiveness. They attributed what William Cullen Bryant had called picturesque qualities to the least majestic portions of the national shore. Although neither man suggested the protection of marshes, each represents a significant departure in the ambivalent American characterization of the shore. Virtually a century would pass before any ecological connection between marshes and estuaries could be understood. After three centuries of discovery, settlement, and economic development, a fonder fascination for marshes had emerged thanks to the romantic movement.[56]

The art, literature, and science practiced by the naturalists established the intellectual basis for society's later acceptance of an estuarine preservation ideal. The negative views of the seashore were countered by a positive feeling that tidelands were beautiful and perhaps necessary to sustain large numbers of fish and birds. In nineteenth-century America, an emergent fondness for the beauty of wildlife and landscape encouraged many seaside visitors to develop a more appreciative and sophisticated ambivalence concerning the utility of recreation and the necessity of tidal marshes.

[56] Bryant's *Picturesque America* includes many seaside areas as picturesque, in addition to the spectacular and monumental scenery of the Grand Canyon, Yellowstone, or Yosemite Valley.

4

Commerce and the Public Trust

> The draining of lakes, marshes and other superficial accumulations
> of moisture reduces the water surface of a country. . . . The
> draining of such waters, if carried on upon a large scale, must affect
> both the humidity and temperature of the atmosphere, and the
> permanent supply of water for extensive districts.
> —George Perkins Marsh, 1864

NATIONAL sympathy for protecting nature dramatically appeared among elite professionals before the Civil War. These preservation advocates conceived the intellectual and tactical foundations of future federal resource-protection policies—especially for fisheries, wildlife, land, and water. During the final decade of the antebellum period reformers focused on state-supported fish propagation and locally created and funded urban parks. During the Civil War this preservationist impulse expanded to include two further objectives. The first was the creation of state parks in order to protect certain extraordinary scenery, such as Yosemite Valley and the Mariposa Grove of giant sequoias in California. The second objective, to restore the natural environment, was broader in scope and became the intellectual basis of the twentieth-century conservation movement.[1]

In 1864 George Perkins Marsh, a linguist, diplomat, and statesman, began to argue that technological society could be viewed as an agent of geologic change—just as glaciers and volcanoes were. He felt that recent fish, bird, and forest depletion resulted from increased economic activity, and he called for a "restoration of disturbed harmonies" in the form of "geographical regeneration."[2]

By the close of the nineteenth century, Marsh's followers, who repre-

[1] Donald Worster, ed., *American Environmentalism: The Formative Period, 1860–1915*, pp. 1–184.

[2] George Perkins Marsh, *Man and Nature: Or Physical Geography as Modified by Human Action*, ed. David Lowenthal, pp. 35–37.

sented a broad spectrum of professional and partisan backgrounds, had expanded the concept of geographical regeneration to encompass the preservation of historic sites, flood control, timber protection, sanitary engineering, agricultural experimentation, irrigation, city planning, and the reclamation of wetlands.[3] This post–Civil War vanguard succeeded in altering the traditional responses of local, state, and federal governments to the American landscape. Concurrently, reclamation continued to convert estuarine mud flats and marshes into economically useful urban, recreational, or agricultural lands.

The origins of fish and bird preservation can be traced to the increasing professionalization of the biological sciences in the nineteenth century prior to the Darwinian revolution. From their inception, fish and bird protection involved the study and reservation of certain estuarine habitats as the sites of bird rookeries or fish nurseries.

Experimentation in artificial fish propagation and the formulation of a plan to restore the nation's ailing rivers both occurred in the early 1850s. Implementation of these plans, though, was beyond the political powers of the pre-war republic and had to await the Reconstruction period. During this period three significant events influenced the eventual formulation of an estuarine preservation ideal.

The first fish and wildlife refuge was created by the City of Oakland, California, in the Lake Merritt estuary during 1870. Second, in 1871 Congress initiated the creation of the U.S. Fish Commission in a first federal attempt to protect a natural resource from depletion. The last event was John Wesley Powell's 1878 publication of a land-use plan for the desert plateau, based in part on a pre–Civil War concern for river restoration. This report was the earliest comprehensive land-use policy for the West, and it rested on an appreciation of the role water played in the creation of the plateau region's diverse habitats.[4] The importance of Powell's report lay in his adaptation of land and water laws, technology, and institutions to the dry, drought-prone West. He influenced subsequent reclamation policies, provided for comprehensive land surveys and land-use planning, and

---

[3] Lewis Mumford, *The Brown Decades: A Study of the Arts in America, 1865–1895*, pp. 26–48; Mel Scott, *American City Planning*, pp. 1–109; A. Hunter Dupree, *Science in the Federal Government*, pp. 232–55; Joseph M. Petulla, *American Environmental History*, pp. 107–235, 266–85; Charles H. Hosmer, *The Presence of the Past: A History of the Preservation Movement*, pp. 102–52, 237–69.

[4] Ira Gabrielson, *Wildlife Refuges*, pp. 3–7; Dupree, *Science in Federal Government*, pp. 232–37, 238.

professed a political faith in communal water ownership. Powell's work clearly rested on a realistic appraisal of the varied lands in need of comprehensive regional planning.

These precedents reached beyond traditional protection of scenic monumentality as exemplified in the creation of a "people's park" in Yellowstone during 1872. The preservation of both rural and urban park lands had a rich heritage dating to William Penn's plan for the tidewater settlement of Philadelphia in 1681 and included such scenic wonders as Arkansas' Hot Springs in 1832.[5] This tradition included Frederick Law Olmsted's New York Central Park in 1857 and the creation of the Boston Regional Park Commission in 1893. Powell's bold ideas for the settlement of the arid lands also rested partially on calculations of human utility. Powell's view of utility, however, was based on organic values and an appreciation for the limits imposed on social and economic institutions by environmental forces.[6]

The underlying difference in these protectionist philosophies is critical to understanding why widespread estuarine preservation was delayed for another ninety years. Lands reserved for the purposes of public recreation or contemplation were part of the traditional preservationist movement, whereas lands retained for irrigation surveys, reservoir construction, timber reserves, or reclamation were novel federal examples of a utilitarian tradition based on concern over future supplies and demands.[7]

As scenic monumentalism led to the creation of national parks and recreation areas, Powell's dream of publicly supported irrigation encouraged agricultural experimentation, designation of permanent timber reserves, and reclamation of wetlands. Fish protection and propagation, fostered by both traditions, eventually led to the preservation of habitats for biological reasons. This later protection rested on a life-centered appreciation of the role of land and water in the sustenance of healthy wildlife communities. Nineteenth-century fish and bird protection foreshadowed the creation of wildlife refuges and primitive areas in the twentieth century.[8]

---

[5] Hans J. Huth, *Nature and the American: Three Centuries of Changing Attitudes*, p. 9.

[6] Frederick Law Olmsted, "Public Parks and the Enlargement of Towns," *Journal of Social Science* 3 (1871): 1–36.

[7] Roderick Nash, ed., *The American Environment*, pp. 18–46; Stewart Udall, *The Quiet Crisis*, pp. 81–108, 171–84.

[8] Ralph Merton Van Brocklin, "The Movement for the Conservation of Natural Resources in the United States before 1901" (Ph.D. diss., University of Michigan, 1952), pp. 83–118; John F. Reiger, *American Sportsmen and the Origins of Conservation* (New York: Winchester Press, 1975).

The U.S. Fish Commission established in 1871 and the Bureau of Ornithology and Mammalogy founded in 1885 were the earliest bureaus within the federal government to exhibit a rudimentary organic perspective toward the national landscape.

Recognition of the interdependence of plants, animals, and their natural environment was encouraged by Darwin's influence. The invention of the German word *oikology*, (in the English of the day, *oecology*), in 1866 by German biologist Ernst Haeckel was one indication of evolution's effect. Two years after the publication of Marsh's *Man and Nature*, Haeckel defined his categorization of an organism's relationship to the environment as the specialty of oecology. In linking this new word with the older term *economy of nature*, Haeckel described ecology as one of the ten subdivisions of biology, one which depicted "the relations of organisms to the surrounding outer world, to the organic and anorganic [*sic*] conditions of existence."[9] This new ecological recognition and the revolution in biological ideas that it represented posed a latent threat to the federal government's control, use, and protection of the nation's natural resources.

As federal bureaus specialized to address the particular problems of agricultural experimentation, geologic, biological, and irrigation surveys, timber protection, and fish conservation, scientists in the civil service competed among themselves for limited appropriations. Until Theodore Roosevelt's administration, the competing federal agencies lacked a cohesive ideal and broad focus. Emphasis in federal scientific research was always divided between applied techniques for sustaining the production of fish, timber, or farms and theoretical research on the relation between geography and the distribution of certain species. The effects of changes in particular environmental factors, like water or nutrients, on the distribution of dominant species therefore was little understood.

The narrower and more immediate focus of practical research led to unilateral projects without any awareness of their probable effects on related resources. For example, in 1886 Nathaniel Shaler, working for Powell's Geological Survey, suggested a bold plan to drain the seacoast swamps of the eastern United States for the purpose of increasing the availability of agricultural lands close to centers of population. His proposals assumed that agriculture was of paramount importance, since any effect of this reclamation on fisheries was ignored. Scientists who pioneered ecology and

---

[9] Ernst Haeckel, *Generelle Morpologie der Organismen*, I, 8, and II, 253–56, 286–87; Ernst Haeckel, *The History of Creation*, p. 447.

marine sciences were still in the early stages of their research.[10] Addison Verrill, a marine scientist with the Fish Commission, had just distinguished species of shore fishes by their affinities for salt, brackish, or fresh water. The full complexity of coastal fisheries' feeding and reproductive habits was not completely understood with relation to marshes and estuaries.

To appreciate the extent to which the later environmental ideal of fish and wildlife protection diverged from the early federal concerns for coastal defense and promotion of commerce, an historical overview of early nineteenth-century federal policies is necessary.

Between the administrations of John Quincy Adams and Benjamin Harrison, the state and federal levels of government became irreversibly involved in timber reservation, scenic parks, wildlife protection, water conservation, agricultural experimentation, international fishery research, and bureau building, which became the foundations of the twentieth-century progressive conservation movement. In an effort to ground this growing federal bureaucratic interference in the market place on constitutional authority, antebellum politicians relied dually on national security and on the protection and later promotion of commerce for the prosperity of the nation. As early as 1789, Congress authorized navigational aids on the coast, and the first timber reserve for naval stores was created in a west Florida estuary on Santa Rosa Island in 1828.[11]

In 1832, the United States Coast Survey was revived and gained the largest federal expenditure in the field of science in the antebellum period. Much like that for the United States Exploring Expedition, 1838–42, the authority to survey the coast and the Pacific Ocean rested on the constitutional provisions of national security and interstate commerce. While the Fish Commission drew undisputed authorization from the commerce clause, subsequent accretions of federal authority were attacked as violations of the constitution and an affront to the spirit of laissez-faire capitalism. When faced with opposition from the older Jeffersonian tradition of limited government, reformers were hard pressed to justify these new bureaus.[12]

Important decisions by Chief Justice John Marshall provided some of the judicial and constitutional basis for later flood, navigation, and water-

---

[10] Samuel P. Hays, *Conservation and the Gospel of Efficiency: The Progressive Conservation Movement, 1890–1920*, pp. 1–26, 92–121, 161–81.

[11] Nash, *American Environment*, p. xv.

[12] William Stanton, *The Great United States Exploring Expedition of 1838–1842*, pp. 23–25, 33.

quality controls of the nation's river systems. In first asserting federal power over interstate commerce in 1824, his interpretation also created a concurrent or dual-control system of federal commerce powers and state police powers in the regulation of rivers, ports, and coastal matters relevant to navigation. The sweeping powers granted to federal authorities under Marshall's ruling in *Gibbons* v. *Ogden*, 1826, provided the justification for revival of the coast survey, federal aids to navigation, and river improvement. Responsibility for carrying out programs to ensure navigation fell to the Army Corps of Engineers under the Survey Act of 1824. Marshall's ruling defined and strengthened the Corps's authority. A subsequent decision of the Marshall court also established the dual responsibility of state and federal governments in the coastal zone. In the absence of federal legislation, Marshall upheld a Delaware law damming a navigable stream for the benefit of the Blackbird Creek Marsh Company, 1829. By this action the court retained the final power of review in balancing the police powers of the state and federal rights to regulate commerce and protect navigation among the states.[13]

The protection of estuarine wildlife originated in the seventeenth century as a colonial responsibility for regulating fishing and hunting and remained a state responsibility under the new constitution. Marshes lying between the highest and lowest tides of the coast were generally considered common property.

Tidelands were held in trust by the states in order to further the public rights of fishing, hunting, and navigation. The Constitution enumerated federal powers and authority over commerce and navigation within estuaries, up rivers, and along the coasts. During the nineteenth century, this joint responsibility for two closely related components of estuaries, waters and tidelands, strained conservation's effectiveness. The settlement of coastal wetlands and flood plains of rivers had necessitated the development of levee, drainage, and pump techniques to protect lowlands from flooding and to remove excess waters. Increased erosion from the expansion of rural and urban lands silted up rivers and required the dredging of navigational channels, especially for deeper-drafted iron ships and steamers.

In the course of the nineteenth century, the discovery of the necessity

[13] Alfred H. Kelly and Winfred A. Harbison, *The American Constitution: Its Origins and Development*, pp. 293–99; W. S. Holt, *Office of the Chief of Engineers of the Army: Its Non-military History, Activities and Organization*, Service Monograph No. 27 (Baltimore: Institute for Government Research, 1923), p. 5.

for a tidal scour to maintain a silt-free channel to the sea in estuaries prompted British and American engineers to evaluate the expanse of water needed at high tide in order to flush the river mouth clear at the ebb tide. Since the tidal marshes are inundated at high tide, these landscapes are essential to the hydrodynamic functioning of many of the nation's leading harbors. A given ratio between marshes and the depth of the estuary reveals the volume of water necessary to flush the navigational channels free of sand or silt. Often the attempts of upland owners to reclaim mud flats and their adjacent marshes tended to reduce the tidal prism or the amount of water held in the estuary during the apex of the flood tides.[14]

The legal recognition of these engineering findings in the late nineteenth century lagged well behind the technology to alter estuarine hydrography. Under prevailing legal opinions, states exercised responsibility in establishing harbor and bulkhead lines, in marking channels, and in passing sanitary laws. In 1866 a railroad company was enjoined from reclaiming its tidal flats because of the company's failure to obtain reclamation permission from the local board of harbor commissioners. The delineation of state jurisdiction with regard to tidelands was earlier addressed by the Supreme Court in 1842 and again in 1845. The more important decision involved the question of whether an owner of New Jersey shorelands could exclude fishermen from tideland oyster beds adjacent to his upland property. Under the argument of riparian rights, a streamside owner could claim the rights to any water flowing past his upland property. This particular New Jersey owner claimed an exclusive patent derived from colonial times to take oysters from the submerged lands that lay between the average tide marks. The Supreme Court contended that the crown, from whom the states inherited their sovereign rights, had held the tidelands in trust for the public to fish, hunt, and navigate in the tidal creeks and sloughs.[15]

Chief Justice Roger B. Taney wrote that "from the time of the settlement to the present day, the previous habits and usages of the colonists have been respected." By that he meant, "They have been accustomed to enjoy in common . . . the benefits and advantages of navigable waters, for the same purposes, and to the same extent, that they have been used . . .

[14] J. Allen, "Hydraulic Engineering," in *History of Technology*, ed. Joseph Singer, V, 547–50.

[15] U.S. War Department, Army Corps of Engineers, *Shore Control and Port Administration: Investigations of the Status of National, State, and Municipal Authority over Port Affairs*, pp. 25–33.

The Dutch developed the *wipmolen*, or water-lifting wind machine, to aid in drain-ing wetlands. Built in 1746, this windmill near Nantucket, Massachusetts, is an early adaptation of the device to the needs of coastal New England. *Courtesy Historical American Building Survey, Library of Congress Collections (MASS, 10 NANT, 6–), 1935*

Left: Frederick Law Olmsted, shown here about 1864, taught that public parks were essential to healthy urban development. *Courtesy California Historical Society, San Francisco*. *Right*: Elizabeth Caroline Agassiz co-authored, with her step-son Alexander Agassiz, the earliest American scientific guide to seashores, *Seaside Studies in Natural History* (1865). *Photographer: Bradley & Rulofson, SF, courtesy California Historical Society, San Francisco*

Louis Agassiz, the foremost biologist of his era, in 1871 founded a school for marine biological research on an island in Buzzard's Bay, Massachusetts. *Photographer: C. E. Watkins, SF, courtesy California Historical Society, San Francisco*

Dredging to keep estuarine channels clear for navigation and to keep croplands drained and protected from flooding was an important part of America's reclamation policy. This disabled dredge was caught as it worked in the delta of the San Joaquin and Sacramento rivers around 1900. *Photographer: Hagerty Photo, SF, courtesy California Historical Society, San Francisco*

Once a fifteen-million-dollar national industry, oyster fisheries have declined because of saltwater intrusion, pollution, and predation. Here oyster fishing boats are shown massed above deep-water beds in South San Francisco Bay, around the turn of the century. *Courtesy California Historical Society, San Francisco*

Oyster reefs first inspired Karl Mobius, a German mathematician, to develop the concept of the biotic community, which became the central organizing concept in ecology. This reef, photographed about 1890, is on M. B. Moraghan's oyster farm in South San Francisco Bay. *Courtesy California Historical Society, San Francisco*

Ebb tide allows this 1890s oysterman to inspect these oyster fenceposts separating rival fishery claims in a cultured oyster bed, seeded on the adjacent flats in South San Francisco Bay. *Courtesy California Historical Society, San Francisco*

Left: Barrier dunes provide flood and storm protection for coastal residents and enclose broad, shallow bays between the mainland and the open ocean. Here, wild sea oats take root and stabilize the dunes above the normal tidal range of the open sea at Caladesi Island State Park, Florida. *Right*: A cypress tree grows among sabal palms and cordgrass along the Weeki Wachi River mouth in west-central Florida.

The savannah grasslands of the Florida Gulf shore are one of the numerous sources of the Gulf of Mexico's rich shrimp fishery. In the background sabal palms grow on high ground formed by debris, silt, and sand the marsh grasses have trapped.

Rachel Carson wrote that the wind-built dunes of a sandy shore give "a sense of antiquity that is missing from the young rock coast of New England." These barrier dunes are at the mouth of Ten Mile River, Mendocino County, California.

Grasses appear second in a sequence of plants whose extensive root systems stabilize dunes by securing sands from being swept inland by sea winds. This grassy dune rises beside a freshwater pond at the mouth of Ten Mile River, California.

Salicornia, commonly called pickle weed, colonizes upper marshes and provides an excellent habitat for mudflat crabs. Dumbarton Bridge spanning South San Francisco Bay alongside this *Salicornia*-bounded marsh pond, dramatizes the importance of estuarine marshes as the terminus of modern transportation networks. *Courtesy Gerald B. Mooney, San Francisco*

An egret stalks the shallows for fish or shrimp along this marshy bank, lined with cordgrass, of the San Francisco National Wildlife Refuge. *Courtesy Gerald B. Mooney, San Francisco*

Shore birds feed along a mudflat in South San Francisco Bay. Rachel Carson sensed in the shore zone "that intricate fabric of life by which one creature is linked to another and each with his surroundings." *Courtesy Gerald B. Mooney, San Francisco*

An egret perched on a tree trunk inspects the marshes of South San Francisco Bay and perhaps observes the encroachment of civilization, represented by the railroad bridge and train in the background. "As civilization creates cities, builds highways, and drains marshes," wrote Rachel Carson, "it takes away . . . the land that is suitable for wildlife." *Courtesy Gerald B. Mooney, San Francisco*

for centuries." Private riparian owners could not, therefore, extinguish the public trust for purposes of navigation and fishing along tidelands.[16]

In 1845, the court declared that states newly admitted to the Union inherited the same privileges as the original colonies. Those privileges included "the same rights, sovereignty and jurisdiction" over the publicly entrusted tidelands or the shores along and land beneath navigable waters. To ensure that the public rights were paramount to those of the riparian owners of shoreland, the court upheld the police power of the several states to provide unobstructed navigation and fishing. River flood plains, usually beyond the reach of tides, were not to be confused legally with tidelands. Since similar types of vegetation existed along upper reaches of estuaries and their adjacent rivers, it took very accurate surveying to distinguish the river's swamp or overflowed lands from estuarine tidelands. Federal grants of wetlands within the public domain to the western and southern states made this distinction crucial.[17]

The United States' acquisition of Florida brought vast swamps and submerged areas into the public domain. As early as 1835 local officials began trying to reclaim these lowlands. In 1845 they petitioned Congress to grant the state swamp and overflowed lands so long as steps were taken to convert those areas into farmland. Surveys were authorized by the secretary of the treasury from 1846 to 1848, but Congress rejected the findings, which showed draining the swamps to be feasible. Although in 1848 Congress tabled Florida Senator Wescott's bill to grant the Everglades to the state for reclamation, Louisiana was granted all swamp and overflowed public lands within the state, providing those wetlands were reclaimed. The following year a dozen other states, including California and Florida, were granted similar wetlands on the condition that the receipts from their sales be used to construct levees and drain the swamps. Approximately 16 million acres in Florida were sold to private interests under the act, and 2,192,456 acres were claimed by California, most of which bordered the extensive delta of the Sacramento and San Joaquin rivers at the head of San Francisco Bay. The states could rely on either federal survey charts or their own evidence when claiming acreage under the swamp and overflowed land acts. The first federal policy concerning drainage consisted of delegat-

---

[16] Ibid., pp. 29–30.
[17] Francois D. Uzes, *Chaining the Land: A History of Surveying in California*, pp. 125–34.

ing the authority to states while retaining the paramount powers over navigation.[18]

This very reclamation of upriver swamps aggravated the siltation of harbors in two ways. The removal of spillways beside the rivers increased the runoff and therefore the areas subject to flooding downstream. Worsening floods in river cities made earlier levees inadequate. Levees themselves created a second problem, because once the land had been drained the soil compacted and subsided, causing a settling of the levees. As the levees on drained lands decreased in height relative to the flood waters, the bed of the river also grew higher from increased deposition of silt from upland erosion.[19]

These interferences in the hydrodynamic aspects of estuaries were not the only problems that beset conservation of coastal wetlands in the late nineteenth century. A growing anxiety over the decline in fisheries and birds that were tied to coastal wetlands for nutrition, nesting, or breeding promoted the protection of wildlife. Preservation advocates were joined by sanitary engineers and civic groups desiring to clean up the nation's polluted rivers in order to provide cities with adequate drinking water and recreational areas.[20]

An appreciation of the natural losses created by settlement, reclamation, and siltation was noted by a civil engineer, Charles Ellet. Appointed to complete a military survey of the Mississippi River Delta, pursuant to the Swamplands Act of 1850, Ellet was critical of the lack of coordination in what passed for flood protection throughout the Lower Mississippi

[18] Ben Palmer, *Swamp Land Drainage with Special Reference to Minnesota*, University of Minnesota Studies in the Social Sciences, No. 5 (Minneapolis: University of Minnesota, 1915), pp. 17–19; John Ise, *United States Forest Policy*, pp. 46–47, 330; W. W. Robinson, *Land in California*, pp. 184, 191–92; W. H. Bryan, "Review of the Tideland Swindle," *California Land Pamphlets*, Vol. 2, No. 1 (March 6, 1866): 1–10.

[19] For particular troubles involved with reclamation in the Sacramento River Delta, see George Frederick West, ed., *History of Sacramento County, California* (Oakland, Calif.: Thompson & West, 1880), pp. 73–74, 186–88, which quotes the state engineer's report on the pitfalls in the noncomprehensive, individualistic approach to levee construction and drainage. Flooding problems were aggravated by hydraulic mining debris. See Robert Kelly, *Gold Versus Grain* (Glendale, Calif.: Arthur H. Clark, 1959), pp. 92, 217, 242–43; Robert Kelly, "Taming the Sacramento: Hamiltonianism in Action," *Pacific Historical Review* 34 (1965): 21–49.

[20] Calvert Grace, "On the Purification of Polluted Streams," *Journal of the Society of Arts* (June 6, 1856): pp. 505–507; George Waring, *The Sanitary Drainage of Houses and Towns* (New York: Hurd & Houghton, 1878); Joel Tarr and Francis C. McMicheal, "Water and Wastes: A History," *Water Spectrum* (Fall, 1978): pp. 18–25.

River Valley. He correctly recognized that the agricultural habitation of the upper reaches of the watershed was ruining the navigability of the entire river system, as well as aggravating the flood levels, particularly in the lower stretches of the delta. In 1852 he recommended "the creation of great artificial reservoirs, and the increase of the capacity of lakes on distant tributaries, by placing dams across their outlets with apertures sufficient for their uniform discharge." Thus he hoped to regulate the flow of the rivers, "to retain a portion of their discharge above until the floods have subsided below." He had recognized that the removal of trees in the higher elevations and the drainage of swamps in the lower levels of the valley had intensified the floods' destructive capabilities, while aggravating the erosion and consequent siltation of the river bottom. Such siltation posed a problem to the extensive and piecemeal system of levees being constructed along the river's edge to protect the settled floodplains from natural inundations.[21]

Ellet expressly noted, "It is proposed by this process to compensate for the loss of those natural reservoirs which have been and are yet to be destroyed by the levees." He critically suggested that "cut-offs" to shorten the navigational channels of the river exacerbated the swiftness of floodwaters and carried silt farther down the riverbed than had previously been the case. The costs of flood control and reclamation in physical terms were further inflated by Louisiana's hiring of engineers to drain swampland in the delta region in compliance with federal grant requirements. Such expenditures as diking, drainage, and higher levees were futile considering the spread of agricultural and urban civilization northward. Ellet pleaded with his superiors and Congress: "The question of whether she [Louisiana] shall be allowed to stand alone and protect herself unaided, from the difficulties forced upon her by the states above, or be sustained by that government which represents the power of all states, is one of deep interest which must be decided by the justice and humanity of the nation." [22]

He called for a multistate compact to rationalize federal river policy lest the existing laissez-faire approach create further loss of property and life and higher local taxes. His visionary recommendation called for a technological remedy to utilize efficiently the original natural advantages of

---

[21] U.S. Congress, Senate, "Report on the Overflows of the Delta of the Mississippi," prepared by Charles Ellet, Sen. Exec. Doc. No. 2, 32d Cong., 1st sess., January 21, 1852, pp. 13–14.

[22] Ibid., p. 17.

the waterway. In 1853 the Smithsonian Institution published his most influential work, *The Mississippi and Ohio Rivers*.[23]

Ellet's predictions were realized in the Mississippi River system because of that watershed's multistate expanse. Equally disastrous flooding
and siltation threatened the Sacramento Valley in California throughout the
1850s and 1860s. Hydraudlic mining techniques for gold recovery added
enormous amounts of gravel and sand to the Sacramento River and its tributaries. During the 1870s and after, the silt washed out of the Sierras by
mining adversely affected the upper portions of the San Francisco Bay,
Suisun Bay, and the delta system. Unlike the Mississippi Delta, which
empties into the Gulf of Mexico, the San Joaquin and Sacramento rivers
converge to form deltaic overflowed lands at the head of a unique estuarine
system comprised of the brackish waters of Suisun Bay, the Carquinez
Straights, and San Pablo Bay. This water merges with San Francisco Bay,
whose sole outlet to the ocean is through the "Golden Gate," a down-
faulted geologic rift in the coastal hills. The upriver sloughs of the delta
region store fresh water and discharge slowly during the usually arid summer and fall. Due to the seasonality of the Pacific Coast's rainfall, large
amounts of sediment are carried down rivers during the wet winter or
spring flood season. The dry lands of summer, therefore, are often submerged by flood waters in the rainy season, and during particularly heavy
years these swamp and overflowed lands may increase enormously. As an
example, the snowfall in the winter of 1849–50 combined with heavy January rains to inundate the City of Sacramento. Levees were proposed by
the city engineer, C. W. Coote, and construction began from funds provided by a $250,000 tax. By 1852 another severe flood destroyed these ini-
tial efforts.[24]

The incredible rains of the winter of 1861–62 were the worst in California's recorded history. Between November 11, 1861, and January 31,
1862, the gold mining town of Sonora reported 102 inches of rainfall. Although reports may have exaggerated somewhat, it was held that no appre-

[23] H. P. Gambrell, "Charles Ellet," *Dictionary of American Biography*, VI, ed. Allen
Johnson and Dumas Malone (New York: Charles Scribners, 1931), pp. 87–88.

[24] Arthur David Howard, *Evolution of the Landscape of the San Francisco Bay Region*
(Berkeley and Los Angeles: University of California Press, 1972), pp. 5–12, 59–71; William L. Kahrl, ed., *The California Water Atlas* (Sacramento: State of California, 1978), pp.
4–14, 104–11; Wright, *History of Sacramento County*, pp. 73–74; Marvin Brienes, "Sacramento Defies the Rivers: 1850–1878," *California History* 58 (Spring, 1979): 2–18.

ciable let-up in the rains occurred during those months. The December 11 *Daily Union* called it "the most destructive flood . . . ever. . . ." The entire valley was said to have been an inland sea, from Marysville to Tulare. The American River levee was breached in Sacramento, and the capital was again under water. Los Angeles reported twenty-eight days and nights of steady rainfall.[25]

During the previous May, the state legislature had created a special Board of Swampland Commissioners. District engineers reported to the board in January, 1862, that the drainage of upriver sloughs and swamps was contracting the area through which floodwaters normally flowed, thereby aggravating the flood level along the diked lands downstream. Local district engineers concluded that the existing system of levees would have to be raised higher every year to avoid future breaches. This was the only assured plan for keeping the rivers within the confines of the levees. As mining debris was carried downstream, the beds of the rivers between the levees were actually raised higher than the surrounding reclaimed lands. When flooding spilled over the inadequate dikes, vast amounts of silt and gravel buried prosperous wheat farms in the valleys. One immediate result of the 1861–62 deluge was the creation by the state legislature of a Board of City Levee Commissioners for Sacramento. Despite these disasters, the granting of swamplands for reclamation continued because it added improved lands to the tax roles.[26]

Swampland grants in California were further decentralized when the legislature granted remaining overflowed acres to affected counties on April 21, 1866. Pursuant to that grant, State Engineer William Hammond Hall's 1878 report to the legislature criticized the lack of a coordinated construction and maintenance policy for the growing levee system. Hall's conclusions bear striking resemblance to Edmund Ruffin's remarks in his 1861 publication on the Carolina shores and to Ellet's warnings in the 1850s. Hall's California findings echoed Ellet's and Ruffin's distrust of free enterprise to obtain flood control. Hall concluded that "the general works designed for the prevention of overflow in this State generally have failed." Having criticized the swampland policies of the state, he argued that the

[25] Walker A. Tompkins, *Goleta the Goodland* (Goleta, Calif.: American Veterans, 1966), pp. 62–63; Brienes, "Sacramento Defies the Rivers," pp. 13–14.

[26] Wright, *History of Sacramento County*, pp. 74, 185–86.

laws set swamp owners against one another, when cooperation was needed to avert flooding of their bottomlands.[27]

Lack of a coordinated levee construction and maintenance policy stemming, in part, from the haphazard federal, state, and local swampland grant system, was responsible for the destruction of valuable wildlife habitats. Dramatic losses in fish and birds excited public concern to defend wildlife. Concurrently, serious urban sanitation and industrial pollution hazards to drinking water supplies disturbed civic life.

Epidemics of cholera swept crowded urban centers with increasing virulence in 1832, 1849, and 1866. The last two epidemics resulted in the creation of metropolitan boards of public health. As flooding had been recognized as a growing problem by civil engineers, concurrent recognition of the poor drainage, tainted drinking water, and filthy conditions of most cities prompted the sanitary reform movement. It was spearheaded by engineers, doctors, lawyers, and women's charitable groups. This movement attempted to radically improve the city environment and curtail river and harbor pollution. Despite the efforts of sanitary reformers, post–Civil War city growth outpaced antipollution efforts. Water quality continued to deteriorate to such a degree that by the end of the century, national action was taken to prevent the pollution of navigable waterways. State commissions established water purity indices, and new sanitation systems were developed to mitigate the dangers inherent in the privy vault and cesspool systems.[28]

Like the advocates of national flood-control planning, the sanitary reformers realized that individual self-reliance could not deal with the regional extent of organic problems. The sanitary reform movement began on the local and state levels after the 1849 cholera epidemic. Frederick Law Olmsted and George Waring played important roles in this move-

---

[27]Kahrl, *California Water Atlas*, p. 23; Wright, *History of Sacramento County*, pp. 186–87; Edmund Ruffin, *Agricultural, Geological and Descriptive Sketches of Lower North Carolina*, pp. 91, 100, quoted in: Suzanne Fries Liebetrau, "Trailblazers in Ecology: The American Ecological Consciousness, 1850–1864" (Ph.D. diss., University of Michigan, 1973), p. 37.

[28]Wright, *History of Sacramento County*, p. 187; Charles Rosenberg, *The Cholera Years: The U.S. in 1832, 1849, and 1866* (Chicago: University of Chicago Press, 1962), pp. 101–23, 142–45, 175–225; Worster, *American Environmentalism*, pp. 131–61; Tarr and McMicheal, "Water and Wastes," pp. 18–25. Robert Clarke (*Ellen Swallow: The Woman Who Founded Ecology*, p. 148) notes that Stephen Smith founded the American Public Health Association in 1871.

ment. Olmsted, a landscape architect of national repute, became interested in the drainage of wetlands on his small farm on Staten Island, New York, in 1848. He and Calvert Vaux were later responsible for the creation of New York's Central Park. Vaux had been a protégé of Andrew Jackson Downing, a pioneer of landscape architecture in America. During their collaboration on Central Park, Olmsted and Vaux also worked with George Waring, a civil engineer, in the drainage of the park area in midtown Manhattan. Olmsted went on to design parks and do urban planning for cities across the country and ultimately influenced suburban extensions of the city. Waring published *The Sanitary Drainage of Houses and Towns* in 1878 and continued to promulgate urban sanitary engineering.[29]

Both the urban park and sanitation movements advocated communal responsibility under state police powers to protect the environment and health of the nation's citizens. Nevertheless, confusion reigned in the formulation of appropriate public health policies. Within the medical science profession, there were arguments over the reasons for the outbreaks. Popular evangelists blamed God. The wealthy blamed poverty and ignorance for epidemic disease. Institutional provision for clean drinking water was a source of major concern to sanitary reform and public health supporters alike.[30]

Until his death in 1903, Frederick Law Olmsted, Sr., furthered the causes of national sanitary reform and urban planning from New York and Boston to California. A tireless advocate of regional planning and civic renewal through the application of technology, Olmsted recognized the organic constraints within which urban and suburban expansion occurred. Olmsted resigned as the executive secretary of the U.S. Sanitary Commission in August of 1863 to take a position as resident manager on the Mariposa Rancho in California. This once-rich gold mining area had recently been sold by John Charles Fremont to San Francisco interests who, in turn, had resold it to some friends of Olmsted in New York City. While in California, Olmsted influenced the future of three nationally recognized institutions. Asked by the governor of the state, Olmsted prepared a recom-

---

[29] Frederick Law Olmsted, *The Papers of Frederick Law Olmsted*, ed. Charles Capen McLaughlin, I, 7–8, 16–55; Udall, *Quiet Crisis*, pp. 171–84.

[30] Waring, *Sanitary Drainage*, pp. 9–21; Scott, *City Planning*, pp. 11, 13, 17; Rosenberg, *Cholera Years*, pp. 192–215. Two park promoters of the Progressive Era also studied under Olmsted: George Kessler helped on Central Park, and Charles Eliot (son of Harvard's president) was an apprentice of Olmsted until 1886.

mendation on the management of the first public recreation ground ever set aside from the public domain by the federal government, Yosemite Valley, and on the Mariposa Big Trees Grove. His second achievement was to draw up plans for the Berkeley campus of the future University of California in 1864–65. The last significant accomplishment was his recommendation to the City of San Francisco in 1865 on the creation of parks, tree-lined boulevards, and suburban development. Unlike the Yosemite report, which was squelched for political reasons, the report to San Francisco was widely circulated, despite the Municipal Board's rejection of the far-reaching proposal.[31]

Many of Olmsted's remarks concerning the urban citizens' need for cheap transportation and outdoor recreation presaged the later developments of rapid transit and the National Park Service. The grace and style of his landscape architecture attempted to wed utility and beauty with the underlying environmental demands of a particular habitat. He recognized that the strength of cities rested on the services they provided. Utilities, water, and sewage were fundamental services that urbanization demanded. For example, Olmsted advocated construction of individual household drains connected to a central sewer line to discharge the wastes from the new water closets (flush toilets) in affluent homes. "Improvement in our sewer system," he noted, "will add considerably to the comparative advantages of a residence in towns, and especially the . . . suburbs." Technology, for Olmsted, was the instrument of civic rejuvenation.[32]

Some of the numerous problems associated with implementing the new technology were addressed by George Waring, whose articles first appeared in the *Atlantic Monthly*. Waring described strong opposition to publicly financed sanitation engineering growing out of popular disregard for community obligations. Waring met their criticism head on: "There can be no equitable or legal private right," he argued, "whose maintenance endangers the well-being of others." Among sanitary reformers it was an article of faith that, as Waring suggested, "drainage . . . should be . . . carried out by public authority."[33]

---

[31]Olmsted, *Papers*, pp. 25, 28–45.

[32]Olmsted, "Public Parks and Towns," pp. 15–19; Tarr and McMicheal, "Water and Wastes," pp. 19–21; Waring, *Sanitary Drainage*, p. 65; Harrison P. Eddy, "Sewerage and the Drainage of Towns," *Transactions of the American Society of Civil Engineers* 82 (1926): 1125–27.

[33]Waring, *Sanitary Drainage*, pp. 1, 11, 12, 15, 127.

Resting on the police powers of the states, municipal authorities were created in New York, Boston, Philadelphia, and many other cities to bring fresh water into and take waste out of their immediate environs. The ideal was in Waring's words "to secure to every member of the community his full supply of uncontaminated air, and . . . of pure drinking water." But his book also warned of the dangers for tidewater ports inherent in waste removal. Although the regional consequences of these actions were not fully appreciated in the 1870s, by the end of the century water and sewage management would force regional planning on many municipalities. Besides the fact that untreated sewage caused mud flats and marshes to emit noisome odors, Waring was concerned about the deposition of sewage into rivers from which downstream towns took their supply of water. He pointed to the accumulation of sewage in the eddies created by slack tides in estuaries as a major health problem.[34]

The growing problems of river pollution and epidemic outbreaks of typhoid, diphtheria, and cholera led to many erroneous suggestions that "miasma" or swamp gases transmitted these diseases. Well into the present century, advocates of tideland reclamation used the ideal of public health to convert marshes from wastelands to agricultural, residential, or commercial uses.[35] Olmsted and Waring were only two outstanding examples of the sanitary reformers concerned with the quality of urban life.

Both the sanitary reform movement and the supporters of urban parks, then, were a mixed blessing to the furtherance of an estuarine preservation ideal. Engineers did come to recognize the importance of swamps, sloughs, and tidal marshes for the flushing of estuaries and the scouring of navigation channels. Many remedies offered by well-meaning professionals, however, lacked the regional breadth to ensure that one city's waste disposal did not contaminate the drinking water of another. The loss of fish and birds was merely one biological indicator of the ailing water quality of the nation's rivers. Disease, sedimentation, and flooding

[34] Ibid.

[35] For example: "Fever and Ague have been common of late years, in many . . . towns. . . . But as the swamps have become more effectively drained, and the lowlands improved, this disease has gradually disappeared, till its occurence is rarely observed" (Nathaniel S. Prime, *History of Long Island*, p. 45). See also: Waring, *Sanitary Drainage*, pp. 45–46, 53. He quotes the Staten Island Improvement Commission's Report, which Olmsted and his close friend Henry Hobson Richardson, the architect, had helped to write. They both believed in the power of farm drainage to abate malarial outbreaks (Olmsted, *Papers*, pp. 39–40).

were others. In general, all of these problems were addressed in the works of George Perkins Marsh. Marsh understood that man, in his unregulated use of the environment, had upset an organic balance. His suggestions that wildlife habitats be restored through long-range planning were a third major source of the emerging conservation movement.[36]

In his report to the Vermont legislature regarding artificial fish propagation, Marsh articulated to a greater degree than Ellet had the impacts of settlement upon river shores. A former congressman, diplomat, and distinguished scholar, Marsh was unequivocal in his blame, clear in his recommendations, and thorough in his historical research concerning artificial fish propagation or pisciculture. He regretted that no "mere protective legislation . . . would restore the ancient abundance of our public fisheries" and noted that New England's political boundaries, separating river sources from their mouths, negated unilateral state actions to restore sport and commercial fisheries. He noted that man was responsible for the precipitous declines in fish, the pollution of water with urban and industrial waste, the unwise clearing of upland forests, and the destruction of insects upon which many fish species depended. Marsh's recognition of the disastrous effects of unrestrained human activities upon fish populations was a crucial factor in later fish and bird preservation. His biographer has suggested that, unlike his contemporaries, Marsh grasped the entire dynamism of living systems and their dependence on an undisturbed environment. Like Charles Darwin, he was "sensitive to the complex nature of the interrelationships and adjustment between species. . . ." More profoundly, however, especially in his later concern for estuarine habitats, Marsh clearly identified the "active role man played in the transforming of the environment."[37]

His recommendation that the state offer premiums to encourage fish-culture research was developed more fully in his major work published in 1864, *Man and Nature*, subtitled *Physical Geography as Modified by Human Action*. His underlying theme, "geographical regeneration," was to

[36]Walter R. Brown, "The Relative Value of Tidal and Upland Waters in Maintaining Rivers, Estuaries, and Harbors," *Van Nostrand's Engineering Magazine*, September, 1885, pp. 177–97; David Stevenson, *The Principles and Practice of Canal and River Engineering* (Edinburgh: Adam and Charles Black, 1886); pp. 238–47; E. A. Giesler, "The Range of Tides in Rivers and Estuaries," *Journal of the Franklin Institute*, no. 132 (August, 1891): 101–11.

[37]David Lowenthal, *George Perkins Marsh: Versatile Vermonter*, pp. 101, 196, 246–47, 250; George Perkins Marsh, *Report Made Under the Authority of the Legislature of Vermont on the Artificial Propagation of Fish*, pp. 12–15.

become the foundation of the subsequent drive to protect the country's natural resources.

While serving as President Taylor's ambassador to the Ottoman Empire, Marsh took the opportunity to survey the ancient landscapes of the eastern Mediterranean. It became thoroughly obvious to this Vermonter, who had witnessed his own native state's rapid settlement, that the human pressures for water, timber, soil, and other limited natural resources lay at the source of the cyclical rise and decay of ancient civilizations. His self-appointed task in *Man and Nature* was to take a comprehensive view of the relations between society and geography. Consequently, his wide-ranging commentary touched upon the extinction of plants, insects, fish, birds, reptiles, and mammals. In addition to explaining the technological alterations of specific landscapes through drainage, canal-building, and mining, he discussed the benefits of having undisturbed forests, soils, and coastal areas. His chapter on waters dealt extensively with siltation and the consequent destruction of estuaries. Marsh admonished his generation for having altered the "distribution and proportions, if not the essential character of the organisms which inhabit . . . even the waters."[38]

By detailing technological encroachment, he disclosed the uncontrollable results of human engineering along the littoral. Human interference had taken a heavy toll in the coastal zone. Aware of the commercial benefits, Marsh commented that man's engineering had "permanently rescued from tidal overflow . . . tracts of ground extensive enough to constitute valuable additions to his agricultural domain." Not always lauding the efforts of reclamationists, though, he argued that "man's 'improvement' of the soil increases the erosion from its surface" and that such actions "compel the rivers to transport to their mouths the earth derived from that erosion." With a stern logic, he warned that the removal of forest cover in the upper reaches of the watershed, when coupled with river and harbor alterations, increased "the transportation of earthy matter to the sea," thereby "gradually filling up the estuaries." His conclusion was unequivocal: "The destruction of every harbor . . . which receives a considerable river must inevitably take place at no very distant date."[39]

Like Ellet before him, Marsh believed that human interference with natural processes necessitated "physical restoration" to avert the conse-

[38] Marsh, *Man and Nature*, pp. 3–5, 35–37, 281; Lowenthal, *Marsh*, pp. 112, 246–50.
[39] Marsh, *Man and Nature*, pp. 365, 366.

quences that all earlier civilizations had encountered. As part of his purpose in writing *Man and Nature*, Marsh suggested the importance of the restoration of disturbed harmonies and the material improvement of exhausted regions. Because man "as a free moral agent" could act with or against physical laws, it was incumbent upon specialists to familiarize themselves with man's effects on organic and inorganic nature. "Until the circumstances conspire to favor the work of geographical regeneration," Marsh had little hope for human cultural survival in harmony with the environment.

George Marsh's stature in the American intellectual community ensured the lasting importance of his seminal treatise. As a Whig congressman, in 1844, he had been instrumental in the creation of the Smithsonian Institution, realized in 1846, and had helped secure the placement of his protégé, Spencer Fullerton Baird, as its assistant director. He supported the work of the American Fish Culturists Association in its advocacy of state aid to artificial fish breeding as part of his 1857 recommendation to Vermont. He corresponded with Charles Lyell, Charles Eliot Norton, co-editor of the *North American Review*, and Franklin B. Hough, an early commissioner of forestry in the Interior Department. Marsh wrote to Baird just before his death in 1881 concerning the popular impact of *Man and Nature*. "Though it has taught little, it has accomplished its end, which was to draw the attention of better-prepared observers. . . ." He greatly underrated the book's significance. In the next century Lewis Mumford was to call the treatise "the fountainhead of the conservation movement." [40]

For the first time in American letters, the estuary had been understood as part of a vast interrelated and dynamic environmental system. A dramatic event on the Pacific Coast, during the time Marsh was writing the book in Italy, showed the wisdom of his vision. The seasonal and torrential rains along southern California's arid slopes precipitated the virtual destruction of what had once been an estuary. In 1774 Pedro Fages had described the present Goleta Slough as "a great estuary" that "penetrates inland by two separate arms." Fages noted that "the estuary spreads continually . . . forming various swamps and ponds of considerable extent." In the early 1860s, the yearly fires that burned the chapparal-covered mountains had left the slopes exposed at a time when the most devastating rainstorms in a century buffeted the entire state. Aggravated by the exis-

[40] Ibid., pp. xiv–xxvii, 3–4, 36; Lowenthal, *Marsh*.

tence of a grazing economy that had appreciably lowered the water tables of the coastal plain, the rain-soaked mud gave way under the pressure of more porous sandstone above, virtually filling the upper reaches of what has since been called the Goleta Slough.[41]

While many individuals even today attribute the initial "reclamation" of the estuary to natural catastrophe, the siltation that closed the Goleta Slough to ocean navigation was a dramatic demonstration of the same processes that were occurring in the Mississippi River and San Joaquin–Sacramento deltas. Siltation was a major factor in the creation of marshes. Beyond a moderate amount of silt, however, siltation inhibits the migration of fish and buries oyster grounds and fish eggs. In California's case, erosive debris from hydraulic mining choked harbors, covered marshes, and buried farmland. Marsh, Ruffin, and Ellet all recognized the effects that human activity had on waterways. Unregulated reclamation, levee construction and maintenance, mining, timber removal, and agricultural activities adversely affect the physical estuarine environment.

The perceptions held by Marsh, Ruffin, Olmsted, and Ellet formed the core of an organic ideology. This ideology gave rise to governmental policies that rejected the preponderant laissez-faire attitudes toward nineteenth-century resource use. Influenced by the writings of these four, a generation subsequently prodded federal policy makers to revise their conception of a narrow legal jurisdiction over land and water. Marsh's followers, in particular, formulated a basis in law for the enlargement of federal power over natural resources through a wider interpretation of the commerce clause. During his survey for the State of North Carolina, Ruffin recognized the inherent dangers of an individualistic policy for the reclamation of swamplands and recommended a coordinated approach. Marsh, while surveying the problem of Vermont's fisheries, recommended interstate cooperation to revive the rivers' fish populations. His experience with the federal survey of the lower Mississippi led Ellet to propose a national means to regulate river flow through dams and reservoirs. Federal laws were the means to assert national authority in watersheds that were under multistate jurisdictions. The park surveys conducted by Olmsted and others led to a profound recognition of the deleterious influences of civilization on nature.

[41]Pedro Fages, *A Historical, Political and Natural Description of California*, trans. Herbert Ingraham Priestly (Ramona, Calif.: Ballena Press, 1972), pp. 1–30; Tompkins, *Goleta the Goodland*, pp. 62–63.

Marsh's appeal for geographical regeneration significantly altered the attitudes of specialists toward land legislation, river policies, and wildlife conservation laws. Local urban parks and sanitary engineering, state game laws, and national scenic reserves were some of the many derivatives of Marsh's "landscape renewal" proposal. Together with Ruffin's, Olmsted's, and Ellet's recommendations, his views epitomized a substantial shift in national sentiments, legal interpretation, and engineering designs in favor of natural systems. Their ideas reflected a systemic understanding of land as a series of biotic communities, and these ideas appeared before such attitudes were commonplace among scientific specialists.

5

Oceanography and Ecology in the Early Federal Bureaus

Every river is a unit from its source to its mouth. A river is a unit, but its uses are many, and with our present knowledge there can be no excuse for sacrificing one use to another if both can be subserved.

—Gifford Pinchot, 1910

AFTER the War between the States, the doleful predictions of George Perkins Marsh concerning fisheries and estuarine habitats churned in the mind of his confidant and supporter, Spencer Fullerton Baird. As a naturalist and the assistant director of the Smithsonian Institution, Baird knew that sport and commercial fisheries were dependent on anadromous fish. These fishes lived most of their existence in the sea but required the fresh-water river sources to spawn the next generation of salmon, herring, or shad. Often the juveniles of these species lingered in the estuaries, where a rich supply of food nourished the young. During the late 1840s, associations of anglers influenced by sportsman Henry William Herbert and scientist Louis Agassiz shared the concerns of Marsh and Baird over the quality and quantity of American fishes. Their misgivings were matched by those of duck hunting clubs, which purchased private reserves among Chesapeake Bay marshes during the forties to sequester a sporting supply of wildfowl.[1]

The problems faced by oyster fishermen—then engaged in an industry of considerable importance—were complicated by the variations in the laws of different states governing their tidelands, the rights of riparian owners, and the public right to gather shellfish. Oyster wars and pirating of oyster farms by poachers led to significant violence in the Delaware, Chesapeake, and San Francisco bays. In rivers that bordered more than one state,

---

[1]James B. Trefethen, *An American Crusade for Wildlife*, pp. 62, 71–75; "Wholesale Pollution of Rivers," *Forest and Stream*, April 29, 1875, pp. 182–83.

unilateral decisions by one legislature were often ignored or contested by
the neighboring assembly. Despite the obstacles to federal involvement in
fisheries, Baird led the effort to establish the United States Fish Commission in 1871.[2]

The magnitude of Baird's achievements in creation of fish and biological surveys is best appreciated by contrasting them with the state of oceanographic science and government in antebellum America. The central factors in policy making before the war were a strict division between national
and state governments, a growing professionalism among the specialists of
an indigenous scientific community, and the quickening pace of technological innovations. By 1825 the national political tradition rested on
three important foundations: the sanctity of private property and contracts,
the inviolability of individual liberty, and the necessity for limited government. Science in the early republic was characterized by the dominance of
the natural history tradition, a preoccupation with practical results from
scholarly research, and the lack of a clearly identifiable American-trained
and self-regulated professional community.[3]

Quick to utilize the technological advantages of steamboats and railways for navigation and transport, Americans were forced to convert more
coastal wetlands into wharfage and railway yard facilities. The growth of
internal and international markets for American manufactures brought the
Pacific Coast within the confines of the nation's frontiers. Eventually detailed charts of the Atlantic, Gulf, and Pacific shores were completed. A
national clearinghouse for scientific data, the Smithsonian Institution, was
created in 1846; and a new bureau to oversee public lands, the Department
of the Interior, was established in 1849.

In response to internal improvements and international trade, federal
bureaus relied on professional surveyors, engineers, and scientists to formulate and implement military, commercial, and land laws. A favorable
disposition toward expert advice among some of the Jeffersonians was later
eclipsed by executive vetoes of internal-improvement legislation and pen-

[2] John Wennersten, "Almighty Oyster: A Chronicle of Chesapeake Bay," *Oceans* (January, 1980): 26–28, 32, 35; Mitchell Postel, "The Legacy of a Lost Resource: The History of
the Fishing Industry off the San Mateo County Bay Line," *San Mateo County Historical Museum Papers* (May, 1978): 3–16; Mary E. Miller, "The Delaware Oyster Industry," *Delaware
History* 16 (1971): 238–54; David Herbert, ed., *Fish and Fisheries* (London: William Black-
wood and Sons, 1883), pp. ix, 1–30.

[3] Richard Hofstadter, *The American Political Tradition*, p. viii; Sally Gregory Kohlstedt, *The Formation of the American Scientific Community*, pp. 1–77.

urious congressional appropriations. John Quincy Adams attempted to re-vive the Jeffersonian wish to sponsor science and the "useful arts" feder-ally, but his tactics led to a thirty-five-year debate over the role, the cost, and the benefits of scientific advice to a republican form of government. The majority's cherished belief in limited government assured that federal sponsorship of pure and applied science would lag behind European research and the efforts of private philanthropists. State-sponsored sur-veys faced similar budgetary constraints but often were the model for later federal undertakings. Constantly on the defensive when seeking public funding, American scientists turned to private monies or were forced to emphasize the practical application of scientific findings. These arguments stressed the military, commercial, or nationalistic advantages of surveys, exploration, or pure research. Little else would have ensured the successful emergence of legislation from the labyrinth of congressional oversight committees.[4]

The debates over the role of government in aiding science have as their context the cherished belief of the masses and controlling elites in limited government. The sustaining prop of these policies was a strict con-struction of the Constitution. Thus when John Quincy Adams introduced his vigorous program of internal improvements through a broad interpreta-tion of the Constitution, he encountered strong opposition and stirred up a serious controversy. Among numerous far-sighted recommendations, the president suggested the construction of "lighthouses and monuments for the safety of our commerce and mariners."[5]

Coastal fortifications, made painfully necessary by the War of 1812, represented an early federal concern for lands lying within the coastal zone. Although hampered by the lack of expertise within the Army Corps of Engineers, these military installations added significantly to the growth of marine engineering in the young republic. Despite the initial hostility to federal expenditure and bureaucracy, Adams and his supporters provided government aid to science based on the necessary constitutional author-ity—national security and commercial expansion.

The Jacksonians added the necessary popular appeal to reestablish the

[4]George H. Daniels, *American Science in the Age of Jackson*, pp. 38–54; A. Hunter Dupree, *Science in the Federal Government*, pp. 57–65, 67–90.

[5]John Quincy Adams, "Inaugural Address," *Inaugural Addresses of the Presidents of the United States: 1789–1969* (Washington, D.C.: Government Printing Office, 1969), pp. 47–52; Dupree, *Science in Federal Government*, pp. 35–43.

U.S. Coast and Geodetic Survey in 1832 and, after interminable delays, to launch the United States Exploring Expedition in 1838. Their desire to "democratize" science and to include nationalism in the debate gained support for domestic and overseas exploration. At the same time, the establishment of a civilian agency assigned the responsibility of the hydrographic and geodetic surveying of the shore sparked lasting resentment from the navy.[6]

The tangible results of the coast survey, especially after a change in directorship in 1842, stand in stark contrast to the mixed blessings of the U.S. Exploring Expedition. Alexander Dallas Bache took charge of the civilian operation in that year and kept it from being reabsorbed by the navy in the early fifties. He immediately began to publish the survey's findings and extended surveying operations beyond the greater New York environs for both scientific and practical political reasons. Besides discovering an easier entrance to New York Harbor, he drew up charts of Boston Harbor—ahead of schedule, thanks to an appropriational inducement from the Massachusetts legislature. Bache plotted the course of the Gulf Stream and employed his colleague and friend Louis Agassiz to study the coral reef formations of the south Florida coast. A master of efficiency and field work and a skilled manipulator of congressional purse strings, Bache had managed to increase the survey's appropriations to a half-million dollars in 1866. The success of General Winfield Scott's strategic blockade of southern ports during the war depended in large measure on the groundwork and baseline studies established under Bache's directorship.[7]

Congressional appropriations for scientific purposes fueled the military-civilian rivalry so characteristic of federal sponsorship. This rivalry was exacerbated once the navy had established the Depot of Charts and Instruments in 1842. Without specific congressional approval, the navy also built the Naval Observatory in Washington for testing the accuracy of its chronometers and to serve as a national repository of celestial data, sailing logs, and charts.[8]

In 1842 Lieutenant Matthew Fontaine Maury succeeded Charles

[6] Dupree, *Science in Federal Government*, pp. 29–33, 44–55, 60–61; Daniels, *Science in Age of Jackson*, pp. 47, 205; William Stanton, *The Great United States Exploring Expedition of 1838–1842*, pp. 3–91, 298–304, 349–77.

[7] Merle Odgers, *Alexander Dallas Bache: Scientist and Educator*, pp. 142–62, 165–73; Gustavus A. Weber, *The Coast and Geodetic Survey: Its History, Activities and Organization*, Institute for Government Research, Service Monographs No. 16, pp. 2–7.

[8] Dupree, *Science in Federal Government*, pp. 62–64.

Wilkes, leader of the U.S. Exploring Expedition, as director of the depot. The young native Virginian set himself to the expansion of the newly designated Naval Observatory and Hydrographic Office in 1843. He soon became involved in meteorological observations and recording the relationship of prevailing winds to ocean currents. After the publication of his "Wind and Current Chart of the North Atlantic" in 1847, he issued a more extensive global map in 1851. Maury's charts reduced the sailing time from New York to Rio de Janeiro by ten to fifteen days and from New York to San Francisco by up to forty-seven days. His interest in currents led to a jurisdictional dispute with Bache over the charting of the Gulf Stream, and his pressure to create national and international weather-monitoring stations ruffled the ambitions of perhaps the most notable native-born scientist of the era, Joseph Henry.[9]

Maury used data from American merchant marine fleets to update his charts and in 1855 published what is considered the first modern textbook on oceanography. The work was entitled *The Physical Geography of the Sea*, and with it he became the first person to treat the ocean as a single discipline and a unified habitat. He extolled his subject, writing: "The study of the sea, in its physical aspect, is truly sublime. It elevates the mind and ennobles man." Although filled with erroneous hypotheses and absurd conjectures, the book has remained a valuable classic because of Maury's sweeping ability to invest the subject with his integrated vision of delicate earthly balances. Those who study the sea, he admonished, "must cease to regard it as a waste of waters." Instead, Maury urged, they "must look upon it as part of the exquisite machinery by which the harmonies of nature are preserved." Two important themes dominate the work: the first, that all creatures were the agents of divine economy, and the second, that there was a life-sustaining connection between the atmosphere and the hydrosphere. He wrote: "The atmosphere is something more than a shoreless ocean. . . . It is an envelope for the dispersion of light and heat over the surface of the earth; it is a sewer into which . . . we cast vast quantities of dead animal matter; it is a laboratory for purification . . . it is a machine for the pumping of all the rivers from the sea, and conveying the waters from their mouths to their sources." In these two assumptions, Maury reflected romanticism's influence through German nature theology as expressed in the writings of Goethe and Humboldt.[10]

[9] Frances L. Williams, *Matthew Fontaine Maury: Scientist of the Sea*, pp. 158–268.
[10] Matthew Fontaine Maury, *The Physical Geography of the Sea*, pp. 53–54, 68.

The value of the pioneering research that Maury and Bache promoted, regardless of their mutual antipathy and rivalry, lay in their establishing the bases for further scientific inquiries into the nature of the seashore. Bache had precisely located the estuaries of the nation; fathomed their channels; identified their shoals, muds, sands, and gravels; published tide tables; and generally rounded out their hydrographic dimensions. Maury, while more concerned with weather and its relation to the Gulf Stream and other ocean currents, grasped the interrelatedness of marine conditions, lending cohesion to the emergent discipline of physical oceanography.

By elaborating the geographical and biological relationships among seashore life and fisheries, scientists contributed to a profound shift in political thought. The interplay of scientists, engineers, and politicians resulted in a practical program furthering both protection of fish and birds and comprehensive river conservation. The ecological concepts of niche, zonation, biophysical regions, and the dependence of species on healthy biotic communities provided intellectual support for an organic political ideology. Against a backdrop of increasing federal engineering patronage, these ecological findings were used by engineers and politicians to counter the prevailing laissez-faire assumptions of the government's limited role.

Once the measures for federal support for science were passed, the changes taking place within the scientific world still affected the kind of expertise available to government. The continued spread of specialties within the older discipline of natural philosophy had by the 1850s replaced the insights of naturalists with the more precise findings of the botanist, zoologist, or geologist. Specialists tended to view the older generalist tradition as passé, and, when the tradition persisted in government circles, generalists' findings were usually discounted. The quest for professionalism within the American scientific community exacerbated this antipathy, and when the National Academy of Sciences was first established in 1863 Spencer Fullerton Baird was conspicuously absent from the membership.

Specialization also affected the study of nature and the formulation of new disciplines within science. Natural history training, as offered in the older colleges, fractured into geology, botany, and zoology in the new scientific schools. Other important information concerning habitats was discovered by physicists and chemists, but interdisciplinary study lagged well behind in professionalization. In government science, naturalists occupied key positions in the early expeditions and later at the Smithsonian Institution. The study of seashores and estuaries was influenced by geologists, hydrographers, marine naturalists, and engineers.

The origins of habitat science reflected the specialties of those early students of the coastal areas. James D. Dana described marine animals' geographic distribution by temperature when he wrote the Wilkes expedition report on crustaceans. Manx naturalist Edward Forbes in 1844 first associated distinct assemblages of sea creatures with specific areas of the ocean—in his terms, "provinces of depth." This fundamental connection between particular animals and their habitat was essential to later descriptions of the niches that organisms filled in the ecology of the sea.[11]

The term *niche*, as it was used by the naturalists, referred to the routine place of residence and the occupation, so to speak, of an organism. The term *nature's economy* was subsequently used to describe the feeding relationships and population changes among species. Niche was derived from the French word meaning nest. Thus, the geographical location or habitat within which an organism found food and shelter or reproduced its young defined only one aspect of a creature's life cycle. The role that the creature played within the biotic community of interdependent plants and animals was also important. A creature's role determined, to a large degree, the ability of a species to acquire food and shelter and to reproduce. The estuaries of any river normally provide more niches for more different types of creatures than the freshwater or adjacent dry-land habitats. When dealing with water habitats, the nineteenth-century scientists had to invent a new nomenclature to discuss feeding, homing, and recreative functions of fishes, crustaceans, algae, or molluscs.[12]

Scientists described marine, estuarine, and river habitats in terms of temperature, depth, salinity, and substrate. James D. Dana, Edward Forbes, and Philip Henry Gosse studied seashore animals, and their findings were popularized by Alexander Agassiz and Caroline Agassiz. Each emphasized different constraints when describing the niche of shore animals. Dana, the geologist, was concerned with temperature and submerged landforms, while Forbes, a naturalist, concluded that the relation to depth or

[11] Margaret Deacon, *Scientists and the Sea, 1650–1900: A Study of Marine Science*, pp. 21–24, 71–75; Justin F. Denzel, "Edward Forbes and the Birth of Marine Ecology," *Sea Frontiers* 22 (January–February, 1976): 16–23; William A. Herdman, *The Founders of Oceanography and Their Work: An Introduction to the Science of the Sea*, pp. 1–60.

[12] Eugene P. Odum, *Fundamentals of Ecology*, pp. 234–39.

pressure was of prime importance. Despite these differences in emphases resulting from their particular methodologies, the isolation of distint faunal groups served as a guide to later studies of estuarine species and the classification of brackish plants and animals.

Forbes's initial success rested on his 1841 publication, *A History of the British Starfishes*, which brought him international acclaim as a marine naturalist. After similar studies of the Aegean Sea, Forbes returned to England, and in 1850, two years before Dana's geographic distribution appeared in the *American Journal of Science*, Forbes described eight zones of life. Of significance for later studies of the seashore was his grouping of organisms found in the intertidal areas, the littoral zone. His untimely death at the age of 39 occurred in the same year as the publication of a book by a fellow Britisher describing seashore biology. Philip Henry Gosse, who also constructed the first aquarium for preserving living creatures, depicted the plants and animals of a particular portion of the Devonshire shore in his 1854 publication. The basis for further systematic study and classification of sea life and sea shores had finally been laid.[13]

In the United States the scientist most responsible for shedding scientific light on marine and freshwater fish and for familiarizing students with European marine biological work was Louis Agassiz. This gifted Swiss immigrant had studied under Cuvier in France and Alexander von Humboldt. A tireless public speaker and fund raiser, he became a Harvard professor in 1848 and lectured to the newly created American Academy for the Advancement of Science on his ichthyological findings from a tour of Lake Superior in the same year. No other single figure in antebellum America did as much as Agassiz to popularize the study of native natural history and particularly the classification of fishes. His professional contacts were impressive: Joseph Henry, the newly appointed director of the Smithsonian Institution; Dana, the recently chosen editor of the *Journal of Science*; Spencer Fullerton Baird; and most importantly Bache, with whom he helped to establish a professional scientific clique called the Lazzaroni. He corresponded with British geologist Sir Charles Lyell, and offered— though he never completed the task—to write up Wilkes's volume on the ichthyological findings of the South Sea expedition. Among his students at Harvard were later scientific notables Addison E. Verrill, William James, Nathaniel Southgate Shaler, and Joseph Le Conte. Thanks to his efforts at

[13] Joel Hedgpeth, ed., *Between Pacific Tides*, rev. ed., pp. vi–vii, ix, 321–514.

fund raising, the Museum of Comparative Zoology at Harvard University was built beside the estuary of the Charles River. His second wife, Caroline, and his son, Alex, popularized seashore nature study. Agassiz's primary achievement in 1873, the last year of his vigorous life, was the opening of the first marine biological summer school on Penikese Island, within Buzzard's Bay, Massachusetts.[14]

Through the efforts of these men of distinction, nature appreciation flourished throughout the 1850s. Louis Agassiz left his son, Alexander, with the responsibility of keeping the Anderson School of Marine Studies operating after his death. While Agassiz has perhaps been known best for his glacial theory and rejection of Darwinian evolution, his concept of plant and animal assemblages as communities was a strikingly modern contribution.

Louis Agassiz's finest exposition of an animal community was of the coral reefs fringing the Florida coast. He had explored the region in 1851 with Alexander. He used the word community in reference to colonies of hydroids. He also noted how mangrove trees take root in the marine humus that sometimes gathers on top of coral reefs in the Florida Keys. To Louis Agassiz, creation reflected the sophisticated mind of the Creator, and the tiny marine polyps covering shells, rocks, or logs in seawater were an example of an interdependent community of organisms. Although Louis Agassiz was an expert on fossil and contemporary fish, in the field of oceanography Agassiz's greatest gift was his son, Alexander.[15]

Alexander became the foremost advocate of oceanographic studies in the nation prior to the First World War. For his philanthropy and study of marine biology, the younger Agassiz has been called one of the American founders of oceanography. He and his father did for fishes in the nineteenth century what the Bartrams had earlier done for the natural history of plants. Alexander received his training at Harvard and its Lawrence Scientific School, in addition to his exposure to Louis's friends. He had accompanied his father to the Florida Keys to study the coral reef formations for the coast survey in 1853. He and his stepmother published the first seaside

[14]Edward Lurie, *Louis Agassiz: A Life in Science*, pp. 1–9, 122–209, 234–51, 341–49, 370–81. See also Herdman, *Founders of Oceanography*, pp. 104–106; Edward Lurie, *Nature and the American Mind*.

[15]Louis Agassiz, "Methods of Study in Natural History," *Atlantic Monthly*, June, 1862, pp. 754–55. Hydroids are one stage in the life cycle of jellyfish, when they are non-floating colonies of individual organisms attached to some stationary thing.

study of natural history in the country, helping to introduce the findings of Gosse and Forbes to American readers. It was not surprising that Alexander was hired by Bache and joined the coast survey in San Francisco during the summer of 1859.

The younger Agassiz, with his stepmother, published descriptions of marine life along the American littoral. Their study of Boston Bay treated the biological life of the seashore as occupying distinct bands or zones, in addition to providing a taxonomic listing of the varied phyla. The work, published in 1865, was entitled *Seaside Studies in Natural History*. The authors responded to a "want often expressed" for a "book of popular character . . . describing the marine animals common to our shores." One major inhibitor of an appreciation for estuarine life lay in the ignorance concerning coastal marine biology, and *Seaside Studies* began the national familiarization with seashore ecology. The final chapter of the book, "Distribution of Life in the Ocean," is of particular importance. The broad habitats of rocky shore, sandy beach, and mud flat were associated with particular assemblages of plants and animals.[16]

In *Seaside Studies*, Alexander Agassiz depicted the Massachusetts shore as a microcosm of the world's oceansides. "A sandy beach, however small," he suggested, "gives us some idea of the nature of animals we may look for on any sandy coast. . . ." In contrast to the rocky shores and sand flats, he noted, "the narrowest mud flat on the seashore" looked "dead and inanimate." But these areas "where the gasses always generated in mud are deadly to many kinds of animals" gave rise to "a variety of worms, who burrow their way into the mud"—as he said—"to court the miasma." Like many of his contemporaries and in the tradition of Dr. Rush, Alexander viewed the mud flats as noxious habitats for all but the most primitive of creatures.[17]

Influenced by the work of Alexander von Humboldt, who had divided the globe into frigid, temperate, and torrid zones, Agassiz referred to the "geographical distribution of animals and plants" as "organic realms." Ecologists now refer to these as biophysical regions, and Agassiz extended the concept to the seashore. Although, he reasoned, "it would seem . . . as

[16] Elizabeth C. Agassiz and Alexander A. Agassiz, *Seaside Studies in Natural History*, p. v; G. R. Agassiz, ed., *Letters and Recollections of Alexander A. Agassiz, with a Sketch of His Life and Work*, pp. 14, 18–23, 25, 44.

[17] Agassiz and Agassiz, *Seaside Studies*, p. 141.

if the water, from its own nature, could hardly impose a barrier as impass-
able as the land," the sea was divided into zones. Agassiz thus concluded,
"The localization of the marine faunae and florae is nevertheless as distinct
as that of terrestrial animals and plants." Of most significance to modern
thought was the way Agassiz attributed this distribution of seashore life to
geophysical conditions. He proceeded to elaborate on the physiographic
characteristics of communities from Baffin Bay to the Gulf of Mexico and
on the Pacific Coast from Alaska to Cabo San Lucas.[18]

Agassiz's discussion of different substrates noted that "the animal life
changed, as we pass from a rockbound to a sandy coast" and "the vegetation also." With some condescension he remarked, "Sea-weeds of the
rocky shore disappear almost entirely, and their place is but poorly sup-
plied by the long eel-grass." While the seaweeds clinging to the rocky
shores are part of the algae family, the eelgrass associated with sand or
mud flats is actually a flowering plant, one of a few in the marine world.
Proceeding down the coast from Cape Cod to Cape Hatteras, Agassiz commented on the effects "the large amount of fresh water poured into the sea
along its whole line" had on "the character of the shore animals." He reasoned, "The large estuaries . . . give a very peculiar character to the
shore, and bring down . . . a large quantity of detritus of all sorts from the
land."[19] Detritus, an important food source for clams, mussels, barnacles,
oysters, shrimp, or insects, is the product of decay—that is, small products
of the disintegration of either minerals or living matter that form an important link between the inanimate habitat and the organic community.[20]

The nation's tropical seas were of special interest to Alexander, as
they had been to his father. Agassiz declared, "Nowhere is there a more
rich and varied littoral fauna" or "a more favorable combination of circumstances for the development of marine life . . . than about the coral
reefs of Florida." In addition to giving the first description of the breadth of
life along the North American shore, Agassiz's work pioneered the application of the notion of life zones related to depth to particular areas of the
seaside. For the sake of clarity, he described seashore zonation in reference
to rocky shores, although he noted that tides and depth influence the biology of sandy beaches and marshes as well. "Along the shores," Agassiz

[18] Ibid., pp. 142–43.
[19] Ibid., pp. 144, 145.
[20] Odum, *Fundamentals of Ecology*, pp. 9, 63.

commented, "animal and vegetable life seems to be distributed in certain definite combinations." He divided "given distances between the high and low water marks" on the rocks into four separate zones.[21]

Although the boundaries he cited are not currently adhered to, Agassiz's division of the tidal seashore is particularly modern. He began with the uppermost areas covered by the high tides, dominated by the rock-weed Fucus, and associated with "barnacles and small Crustacea." Below this assemblage were certain algae and "Star-fishes, Crabs and Cockles." His designation of the final two zones reflected the influence of Edward Forbes. In describing the lower tidal zones Agassiz wrote, "Next in order is the Laminarian . . . home of the Sea-urchin" and "lastly we have the Coralline zone." This final area below the tidal ebb contained lobsters and numerous clusters of hydroids, the asexual phase of the jellyfishes' life cycle. Agassiz summarized seashore zonation, asserting, "This distribution is not casual; these belts of animal and vegetable life are sharply defined and so constantly associated, that they must be controlled by the same physical laws."[22]

Agassiz provided the nation with its first account of biological communities, tidal zonation, and biophysical regions of the seashore. On a tour of Europe in the summer of 1870, Agassiz had occasion to meet and talk with Ernst Haeckel. After their meeting, he observed, "Haeckel I liked extremely," but "he has left the positive for the speculative and indulges in fancies. . . ."[23]

Although the philanthropic support for the summer biological station on Penikese Island was withdrawn and the Anderson School closed in the seventies, Agassiz's subsequent career prospered. He conversed with Charles Darwin, Prince Albert of Monaco, and Sir John Murray, the director of the century's greatest oceanographic undertaking, the Challenger Expedition (1872–76). These men shared Agassiz's interest in marine science. Grover Cleveland failed to convince Agassiz to take on the post of scientific adviser to the president in 1885. From his shoreside home in Newport, Rhode Island, Agassiz continued to sample and study shore life,

[21] Agassiz and Agassiz, *Seaside Studies*, pp. 148–49.

[22] Ibid. For a modern discussion of the various ways to divide intertidal zones and their confusing nomenclature, see Joel Hedgpeth, "Classification of Marine Environments," in *Treatise on Marine Ecology and Paleoecology*, vol. 2 in *The Geological Society of America*, Memoir No. 67, ed. Harry S. Ladd (Washington, D.C.: National Research Council, 1957), pp. 93–99.

[23] Agassiz, *Letters of Alexander Agassiz*, p. 108.

though his major concerns were deep-sea or pelagic marine life. Through the careers of both Louis and Alexander Agassiz, the national interest in fish and fisheries was furthered and popularized. They reflected the growing interest in the life histories of fish by amateur and professional scientists as well as anglers and sportsmen.[24]

Thanks to the contributions of Maury, Bache, Dana, and the Agassizes, the national fascination with the seashore was coming to rest on a scientific appreciation for coastal wildlife. While not totally replacing an inherited ambivalence toward nature, the growing cultural regard for estaurine shores was reflected in the action of sportsmen and citizens alike. By the 1840s, duck clubs in the Northeast were purchasing wild portions of the Chesapeake Bay as hunting preserves. In 1870, the city of Oakland, California, set aside Lake Merritt (originally a tidal basin) for the purposes of fish and bird protection. While these scattered examples did not give rise to a mature estuarine preservation ideal, they did reflect the beginning of a change in the national temper regarding wild shores.[25]

The contributions that scientists of this period made to an understanding of estuarine habitats can be appreciated only in retrospect. No national program singled out river mouths for biological study. Estuaries were mapped as part of the coast survey, hydrographic data were obtained by engineers for bridge, pier, or jetty construction, and some biological studies were conducted by states or local societies. The early contributions of marine research established the wide parameters within which estuarine ecology subsequently developed. Life zones in the sea were delineated by Edward Forbes, Charles Dana, and Alexander Agassiz. The specific relationships of organisms within those zones remained for later investigation. Cross-disciplinary oceanography and ecology relied on the growing specialization of associated sciences for accurate data. The slower professionalization of marine and ecological studies compared with that of chemistry and physics delayed a comprehensive knowledge of estuarine biology. Only studies of bird behavior and fisheries formed the basis from which modern ecologists have built their discipline.[26]

[24] Ibid., p. 219; Herdman, *Founders of Oceanography*, pp. 107–18.

[25] Trefethen, *American Crusade for Wildlife*, p. 72; Lurie, *Nature and the American Mind*, pp. 379–81.

[26] Warder C. Allee et al., *Principles of Animal Ecology*, pp. 13–21, 731–801; Richard Brewer, "A Brief History of Ecology: Pre-nineteenth Century to 1919," *Occasional Papers of the C. C. Adams Center for Ecological Studies* 1 (November 22, 1969), Western Michigan University, Kalamazoo, pp. 1–17.

A consequence of the specialization and slower growth of marine science was that engineers who studied natural contours were the first to recognize estuaries as dynamic entities, which transported sediment, liquid and solid wastes, and deeper-draughted shipping. This was one ironic aspect of nineteenth-century estuarine research. Wildlife protectionists, often motivated by a biocentric system of values, explored individual biotic communities within the estuarine habitat. Although these organisms were often indicators of the condition of the water body, engineers motivated by utilitarian concerns for estuarine projects were the first to explain the river mouth as an integrally dynamic unit. Partly because of this earlier recognition, civil engineers were particularly influential in establishing federal policies concerning the nation's rivers. Their reclamation programs, resting on the legal grounds of maintaining interstate commerce, thus overshadowed and set back the more preservation-oriented approach of the ecologists.[27]

While federal efforts to restore the nation's rivers were heavily influenced by the engineering profession, the older legacy of the naturalists was kept alive and flourished in federal bureaus thanks to two extraordinary individuals. One was Spencer Fullerton Baird, the protégé of both Marsh and Audubon, whose position at the Smithsonian Institution gave him the experience needed to coax federal money for fish and bird studies. Baird was joined by John Wesley Powell in the institution's Bureau for Ethnography. From these relatively secure positions in the federal scientific establishment, Baird created the United States Fish Commission, and Powell guided the U.S. Geological Survey. Baird established the first national program for the restoration and conservation of wildlife, and Powell struggled in vain to make reclamation of arid lands a national priority.[28]

Baird was instrumental in the creation of the U.S. Biological Survey and the appointment of ethnographer-naturalist C. Hart Merriam as its director. These figures took weak bureaus and forged the institutional basis for wildlife preservation and river conservation during the later Progres-

[27] Allen, "Hydraulic Engineering," in *History of Technology*, ed. Joseph Singer, V, 547–51; Eugene Weber, "Comprehensive River Basin Planning: Development of a Concept," *Journal of Soil and Water Conservation* (July–August, 1964): 133–38. On Reynolds's influence in America, see E. A. Giesler, "The Range of Tides in Rivers and Estuaries," *Journal of the Franklin Institute of Philadelphia* (August, 1891): 101–11.

[28] Dean Conrad Allard, Jr., *Spencer Fullerton Baird and the United States Fish Commission*, pp. 1–65; William Healey Dall, *Spencer Fullerton Baird: A Biography*; William Culp Darrah, *Powell of the Colorado*, pp. 222–340; Wallace Stegner, *Beyond the Hundredth Meridian*.

sive era. They understood the relationships among water, landscape, and social institutions. They meshed the concerns of nature enthusiasts with realistic programs to further George P. Marsh's ideal of "geographical regeneration." Quite unintentionally they also added a new dimension to an estuarine preservation ideal. Their advocacy of land-use planning for watersheds or habitat protection for endangered species furthered the broader supports on which estuary protection eventually rested. The changes these people facilitated can be traced to a national concern for the decline in fishes alluded to earlier by Marsh in Vermont and others in Massachusetts.

By the summer of 1868 Baird had become a seasonal visitor of the shores of Buzzard's Bay, Massachusetts, and as a curious scientist he was interested in the marine life of the estuary and its shores. He was aware of a conflict that had arisen between the coastal fishermen of adjacent Rhode Island and Massachusetts. Line fishermen blamed the traps used by boat fishermen for the noticeable decline in the commercially important species of sea bass, striped bass, and scup. When the state authorities involved in determining the cause of their demise came to differing conclusions, each state recommended different laws for regulating the fishery. Baird seized this opportunity to get Congress to underwrite a national commission much like the one New Hampshire had first set up in 1864. Baird's drive was backed by the American Fish Culturists Association, founded in Albany on December 20, 1870.

In the midst of the dispute Baird visited the area of Woods Hole, Massachusetts, with a professor from Union College and, more importantly, George Edmunds, senator from Vermont, Baird's father-in-law. While Massachusetts authorities had recognized the destructive role of man in river or lake fisheries, they had concluded that human damage to the pelagic fisheries was insignificant. Once a fishery-control bill was defeated in the New Hampshire legislature, the U.S. congressman from Massachusetts, Henry L. Dawes, chairman of the House Appropriations Committee, swung his support to the national commission. In January of 1871, the United States Fish Commission was created, and on March 31 it was granted $5,000 to conduct a study of New England's fish and fishery methods. Thus, the first federal action in the realm of marine biology became a reality.[29]

Although Baird was mistaken about the exact reasons for the decline

[29] U.S. Fish Commission, "The First Decade of the U.S. Fish Commission," prepared by George Brown Goode, in *Report of the U.S. Fish Commissioner, 1880*, pp. 53–62. See also Robert H. Connery, *Government Problems in Wildlife Conservation*, pp. 115–43.

of coastal fish, he expanded the work of the commission in 1872 to include a study of an artificial means of restoring the fishery. He hired Addison E. Verrill, Louis Agassiz's former student, to assist with the collections and reports. Significant achievements are traceable to the commission, in light of later developments in estuarine science. Baird had secured permanent support for the United States Fish Commission in 1874. The bill was introduced by Senator James B. Roosevelt of New York, a later president of the Fish Culturists Association and Edmunds's important ally. In the same year, based on his extensive research and collections, Addison Verrill described a unique estuarine grouping of organisms as one of "three quite-distinct assemblages of animal life, which are dependent upon and limited by definite physical conditions of the waters which they inhabit." [30]

Recognizing the estuarine nursery role in fisheries, Baird took issue with the Massachusetts board's earlier findings. He reported that "the river fisheries have been depreciated or destroyed by means of dams or exhaustive fishing." [31] The purpose of his now permanent agency, whose conservative opponents feared it signalled the growth of an irreducible federal bureaucracy, was threefold. First, the agency was to investigate the biological and physical dimensions of the nation's waters. Second, it was to study and publish past and present methods of breeding fish artificially. This study was to facilitate its last and most important directive, the introduction of useful food fish into the nation's river systems.

At the national level, opposition to the commission's work came from conservatives in government, who balked at these new responsibilities of an agency directly under Congress's control. But Baird also identified local opponents in the affected areas. "Unfortunately the lumbering interests in Maine, and the manufacturing interests in New Hampshire and Massachusetts, are so powerful," he wrote, "as to render it extremely difficult to carry out any measures which in any way interfere with their convenience or profit." While singling out these land-based groups for criticism, Baird was not blind to the threats posed by new maritime technologies and processing interests whose livelihoods depended on supplying an ever-

[30] Addison Verrill and S. I. Smith, "Report on the Invertebrate Animals of Vineyard Sound and Adjacent Waters," in *Report of the U.S. Commissioner on Fish and Fisheries, 1871–72* (Washington, D.C.: U.S. Government Printing Office, 1874), pp. 295–352. See also Addison Verrill, "On the Polyps and Echinoderms of New England with Descriptions of New Species," *Proceedings of the Boston Society of Natural History* 10 (1866): 333–57.

[31] Spencer F. Baird, *Report of the United States Fish Commissioner, 1872–73*, pp. xii, 469.

growing food-fish market with cod, shad, herring, or salmon. The practice of ice packing widened the availability of these products to the rapidly expanding urban population of the Northeast. The involvement of federal authorities had come none too soon, for as Governor Cheney of New York later remarked, "About 1880 the shad resorts of the Atlantic Coast were in a deplorable condition" and had to be "restocked by artificial processes." [32]

More than any other single accomplishment of the fish commission, the operation of fish hatcheries and their aid to state programs was an important service. In this manner the nation averted the local extinction of indigenous species that marked bird and mammal populations during the same era. Through the efforts of the fish commission, local depletion of salmon, shad, trout, or oysters could be reversed by the introduction of exotic strains. Pacific salmon were shipped east by specially designed railway cars, and striped bass were taken from the east and introduced in San Francisco Bay. Like the introduction of the English sparrow, the transplanting of exotic species often led to unexpected consequences. For example, the introduction of eastern oysters to San Francisco Bay brought with it an oyster parasite or oyster drill. Having no natural enemies locally, the eastern oyster drills multiplied and became one of three significant factors in raising the price of Calfornia oysters to double that of the East Coast species. However, predation was the primary reason for the decline in San Francisco Bay oysters; the bat stingray and human poachers or oysters pirates helped decrease the oyster harvest. While the fish commission could not police the oyster pirates, it did provide valuable information on the desired water temperature, salinity, and effects of oyster dredges on this nationally important shell fishery. [33]

Two further achievements were of note in the early successes of the bureau. The international acclaim won by the fish commission in the Berlin, Edinburgh, and London fishery exhibitions ranked the country second to none in its pioneering of fish hatchery technology. The yearly reports of the fish commissioner, in addition, brought to local scientists and private fish culturists valuable information concerning domestic and foreign research. None was more significant to the future intellectual development of ecology, or the particular understanding of estuarine science, than the studies of the German mathematician Karl Mobius.

[32] Ibid., p. xi; U.S. Fish Commission, "Proceedings and Papers of the National Fishery Congress," *Bulletin of the U.S. Fish Commission, 1897*, p. 154.

[33] Postel, "Legacy of a Lost Resource," pp. 2–17, 34–48.

As far back as 1838, in the tradition of Humboldt and Goethe, the German geographer August Grisebach at Gottingen had referred to particular associations of plants, such as a forest, grassland, or tundra, as formations. He had demonstrated that plants in differing geographic locations have a similarity of adaptations governed generally, as Humboldt had first noted, by the prevailing climatic regime. Thus the steppes of the Ukraine, Poland, South Africa, or the Great Plains of North America, all tended to support a grassland formation rather than a deciduous forest. Mobius extended this concept of aggregation from terrestrial plants to aquatic animals, and his earliest example was that of oyster formations in the estuaries of his native land. "Every oyster-bed is thus to a certain degree, a community of living beings," he had written, stressing community in contrast to the emphasis placed on competition by the prevailing Darwinian strain of science. Mobius described this community as "a collection of species and an amassing of individuals." Because of the inadequacy of climate in explaining the distribution of marine life, Mobius was forced to consider a variety of factors—"such as suitable soil, sufficient food, the requisite percentage of salt," and "a temperature favorable to their development"—that acted as a matrix assuring the survival of oysters.[34] Beyond the substitution of cooperative development for competitive evolution, Mobius had made a lasting contribution to the nomenclature of that division of biology which Ernst Haeckel had called ecology.

As there was no accurate term to express the reciprocal relations of oyster beds with their environment, Mobius coined the word *biocoenosis*. This term has remained the European equivalent of what American ecologists called the "biotic community." The terms refer to networks of plant and animal species that live in close association with one another. Mobius determined that coral reefs and oyster reefs of estuaries provided a rich habitat for the nourishment of other creatures. Oyster reefs affect sedimentation and tidal flow, but they are in turn hampered by either the intrusion of salt water during a drought or burial by silt during a flood. Since oysters require coarse material to settle upon during the larval stage, siltation can disrupt their process of maturation by burying the suitable substrate with mud during flood times. Mobius noted the adverse affects of environmental changes on oyster cultures. He suggested, "If, at any time, the external

[34] Karl Mobius, "The Oyster and the Oyster Culture," trans. in *Report of the U.S. Fish Commission, 1880*, pp. 683–751; Donald Worster, *Nature's Economy: The Roots of Ecology*, pp. 194–95, 202.

conditions of life should deviate for a long time from the ordinary mean, the entire . . . community would be transformed." [35]

Where Mobius showed how the habitat infringed on the biotic character of an organism, Haeckel, in 1890, used environmental factors in order to distinguish whole groups of sea life. Referring to all of the organisms of the sea as "halobios," Haeckel differentiated these ecologically into "benthos and plankton." The former were either sessile or vagrant bottom-dwelling creatures, while the latter were either free-swimming "nekton" or drifting organisms at the mercy of the currents and tides. He further distinguished the near-shore, or littoral, benthos from the deep-sea, or abyssal, benthos. In describing his methodology, Haeckel emphasized the role of "great oecological differences." [36]

The importance of Haeckel's work, and that of Mobius, is demonstrated by today's continued reliance on their terminology. Both studies—Mobius's 1880 article and Haeckel's 1890 work—were made available to fishery biologists by the U.S. fish commissioner's yearly reports. Consequently, the nation's fishing industry was kept abreast of European oceanographic research.

Baird and his successor, G. Browne Goode, traveled extensively in Europe to attend international fishery exhibitions. They returned to the States with the idea of constructing marine biological research stations modeled on Anton Dohrn's first sustained lab, the Stazione Zoologica at Naples, Italy, established in 1872. Baird created the first U.S. fisheries station at Woods Hole, Massachusetts, in 1882. After his death in 1887, successive commissioners followed this lead at Beaufort, North Carolina, in 1901 and at Key West, Florida, in 1916. Before his career ended, Baird made another decisive contribution to the course of federal science. It was partially as a result of his counsel that the Bureau of Economic Ornithology and Mammalogy was established in the Department of Agriculture in 1885. Baird recommended C. Hart Merriam to be director. Merriam successfully promoted both the movement for bird protection and the intellectual development of the biotic-community concept. [37]

---

[35] Mobius, "The Oyster and Oyster Culture," pp. 683, 688.

[36] Ernst Haeckel, "Plankton Studies" (1890), trans. George Wilton Field in *U.S. Fish Commissioner Report, 1889–1891*, pp. 565.

[37] Herdman, *Founders of Oceanography*, pp. 135; C. Hart Merriam, "Baird, the Naturalist," *Scientific Monthly* 5 (June, 1924): 588; Trefethen, *American Crusade for Wildlife*, pp. 113–14.

The contributions of Merriam to an understanding of biology far out-weighed his later recommendations concerning the federal role in protect-ing wild game. After his field survey of the San Francisco Mountains of Arizona, Merriam devised the concept of life zones, segregating six dis-tinct biotic communities with reference to altitude and temperature. Every mile he ascended a peak was equivalent to 800 miles in latitude so far as the biotic communities were concerned, he concluded. On the very upper levels of the mountain were plants and animals that thrive in the Arctic tundra, while at the base of the peak there existed Sonoran Desert species. With the exception of the tropical zone's biotic communities existing on the southern tip of Florida, Merriam had broadly defined the natural ranges of plants and animals throughout the country. A similar zone concept had been applied to marine biotic communities earlier by Forbes and Agassiz. Together with Merriam's findings, their research popularized the designa-tion of the zones within the biophysical regions into which the coast is now divided.[38]

As a self-trained zoologist, Merriam, a physician by profession, be-came involved with the Bureau of Economic Ornithology and Mammalogy through his participation in the American Ornithologists Union (AOU). The AOU was created in October of 1883, after the Fish Culturists Associa-tion, to lobby Congress for the necessary work of studying the relationship between birds and insects. The helpful role of insectivorous birds had been appreciated since the antebellum period, when the English sparrow had been imported to deal with a caterpillar explosion. However, the introduc-tion of an "exotic" species with no natural predators in this country had precipitated sparrow overpopulation. On October 1, 1884, William Brew-ster, head of the Museum of Comparative Zoology at Harvard, addressed the AOU on the need for bird protection and a study of the sparrow prob-lem. One group of fellows was established to write a model protective stat-ute for non-game birds, while a group headed by Merriam was concerned with determining the range of native birds. The work involved coun-trywide reporting and so overwhelmed the AOU's committee that Merriam asked for congressional help to carry on the survey. With the help of Baird and others, Congress was convinced to appropriate funds in 1885, so that the Department of Agriculture could set up a section of economic ornithol-ogy, within the Division of Entomology.[39]

[38] Worster, *Nature's Economy*, pp. 195–98.

[39] Cameron Jenks, *The Bureau of the Biological Survey*, p. 145; Robert H. Connery, *Governmental Problems in Wildlife Conservation*, pp. 81–97.

The eighties became for bird preservation what the seventies had been for fisheries: the formative period of federal institutional concern. Birds became one of the primary biological indicators of the destructive tendencies inherent in urban, industrial advance throughout the Gilded Age. The use of plumes in the millinery industry gave rise to the highly profitable feather trade and the rather unsporting occupation of the market hunter. The growing demand for bird feathers in women's fashions attested to the growing ranks of middle-class women in late Victorian society. Corresponding declines in waterfowl dismayed sport hunters and the growing number of women's social organizations—not to mention scientists, bird watchers, and state game officials. The birds Audubon shot for sport and art were being exterminated for pecuniary advantage by numerous market hunters with automatic weapons.[40]

In 1886, Merriam, as director of the Bureau of Economic Ornithology and Mammalogy, was in a separate division within the Department of Agriculture. Data on birds' life histories and migratory patterns were collected, along with information on the role of birds as controllers of weeds and insects. Merriam also produced information to aid in the federal protection of plume birds, whose numbers were being rapidly diminished by market hunters. Gulls and terns residing along the shore were particularly threatened by the market hunters, as were herons, egrets, ibises, and roseate spoonbills. These species maintained rookeries in and frequented the numerous sloughs and salt marshes of the Atlantic and Gulf coasts' barrier islands. While state laws prohibited the market hunting of these birds, few game wardens were appointed. Though national in scale, Merriam's division, later called the Biological Survey, lacked police powers.

As the danger to coastal rookeries increased through the 1890s, attempts were made by Congressman John F. Lacey to make a federal offense of the interstate shipment of game birds killed in violation of state laws. Having failed in 1897, the act was finally accepted in 1900, though many doubted its constitutionality.[41] Two government actions and one public organization were partially responsible for this turnabout, with additional support from the AOU, sportsmen, and Women's Clubs. The first action was President Harrison's creation of the first fish and forest preserve on Afognak Island, Alaska, for the purpose of protecting an entire habitat and its biological communities. The 1892 proclamation recognized the inter-

[40] Robin W. Doughty, *Plume Bird and Feather Fashions*, pp. 59, 61–93; Trefethen, *American Crusade for Wildlife*, pp. 114, 117, 120, 129–56.
[41] Trefethen, *American Crusade for Wildlife*, pp. 122–26, 157–221.

relatedness of the plants and animals of the delicate island biology. It noted the island was required for "public purposes," in order that salmon and other fish, sea mammals, birds, and timber "on and about the Island may be protected and preserved unimpaired. . . ."[42]

This unprecedented use of the executive department's powers under the Timber Reserve Act of 1891 represented an early federal protection of both terrestrial and marine geographic habitats and their living communities. This remote place off Kodiak Island, Alaska, was chosen over others for wilderness preservation because any federal attempt to protect wildlife was secondary to encouraging homesteading on public domain lands in the West. Had this island been fit for farming and in one of the public-land states instead of a territory, local opposition in Congress would probably have blocked this use of executive discretion. It had been a backlash by just such western farming and ranching interests that had killed John Wesley Powell's irrigation survey in 1890 and driven Powell out of the Geological Survey in 1894.[43]

Coupled with the failure of Powell's effort to formulate a comprehensive land-use plan for the West, financial depression in 1893 led Congress to cut drastically federal support for all branches of government science. With Baird dead, Merriam, Goode, and others in federal bureaus constantly had to show the economic necessity of their research before the ever-scrutinous eyes of Congress. Although in 1896 Congress recognized the broader scope of Merriam's work by changing the name of his bureau to the U.S. Biological Survey, for the next eleven years his group's broad-based research was under constant threat of appropriations cuts.

Nonetheless, these two government decisions—the executive recognition of the need for habitat protection and the legislative support for a biological survey—were important precedents for the more revolutionary departures in federal resource policies occurring in the Theodore Roosevelt administration. If these federal actions on behalf of wildlife were economically miniscule, the public response to George Bird Grinnell's idea for the creation of state Audubon societies for the promotion of bird protection was overwhelming. Between 1886 and 1889, fifty thousand pledges from members of these Audubon societies not to kill birds or wear the skins of dead birds attested to the widespread popularity of Grinnell's plan.

---

[42] Benjamin Harrison, "Afognak Island Proclamation" (December 24, 1892), in *U.S. Statutes at Large*, vol. 27, no. 39, p. 1052.

[43] Darrah, *Powell of the Colorado*, pp. 310–45; Stegner, *Beyond the Hundredth Meridian*, pp. 235–40.

Thus public support, professional scientific societies, sportsmen's clubs, women's clubs, and federal scientists were all part of the change in mood facilitating the passage of the Lacey Act in 1900. The first federal law protecting wildlife, its enforcement fell to the Biological Survey.[44]

Fish and wildlife protection significantly influenced Progressive conservation through Baird's recruitment of people who later shaped ecological ideas and federal policies. John Wesley Powell as head of the Geologic Survey also guided the intellectual and political development of natural resources regulation by hiring a number of highly qualified professionals. Imbued with Powell's reclamation vision, Lester Ward, Nathaniel Shaler, W. J. McGee, and Frederick Newell articulated new conceptions and fashioned comprehensive policies emphasizing utilitarian values to the exclusion of aesthetic or scientific considerations. Ward and Shaler offered a vision of social cooperation as the pragmatic basis of Progressive reclamation. The cultural improvement of marginal lands through federal intervention to benefit local communities became the paramount conservation objective advocated by Shaler, McGee, and Newell. The policies and tactics for implementing conservation—a term coined by McGee in 1907—centered around comprehensive riverine management, which was intended to revive entire regions, and were derived from the writings of these seminal figures.

A radical interpretation of Darwin's nature studies advocated by Lester Ward stressed communal cooperation for species survival. Drawing also on Alexander von Humboldt's works, the idea of the coevolution of plant and animal combinations was popularized by Ward and Shaler. Trained in geology and a one-time student of Louis Agassiz, Shaler clarified this understanding of Darwin, which was distinct from the prevailing popularizations of the "survival of the fittest." Shaler suggested in 1905 that organic life was "a group of vast associations" of species "united as in a commonwealth."[45] This view was also based on the findings of Mobius and later elaborators of his idea of *biocoenosis*, or biotic community. Shaler was an important transmitter of European ideas and, even after his death in 1906, had a major influence on national reclamation policies.

Shaler was a prolific writer about the seashore, its biotic communities, and its resource potentials, beginning with his first survey of Mount

---

[44]Dupree, *Science in Federal Government*, pp. 232–55; Trefethen, *American Crusade for Wildlife*, pp. 129–38.

[45]Donald Worster, ed., *American Environmentalism: The Formative Period, 1860–1915*, pp. 4–10; Nathaniel S. Shaler, *Man and the Earth*, p. 191.

Desert Island, Maine, for the U.S. Geological Survey in 1886. Later in the same decade he published a report for the Survey on the seacoast swamps of the Atlantic shores in which he called for a national reclamation policy for wetlands. Like Mobius, Shaler demonstrated the impact of external or "edaphic" forces on the development of biotic communities. In his 1886 recommendation for a national reclamation policy for the commonly or privately owned coastal wetlands, Shaler also observed that "no other case in nature" more beautifully demonstrated "the adjustment of organic relations to physical conditions . . . [than] the balance between the growing marshes and the tidal streams." [46] Later, as dean of the Lawrence Scientific School, Shaler wrote a popular work on coastal geology called *Sea and Land*, in which he developed these thoughts concerning the relationship of biotic communities to geographical habitats.

Like others of his time, Shaler noted that the interplay between biological and geological forces tends toward a "balanced state," or equilibrium. This concept of equilibrium was an important idea in the works of both European and American ecologists who tried to show a relationship between transient and permanent plant associations. As the plants changed their conditions over time, so did the biotic community sustained by the new vegetation. [47]

The concept of community within a changing physical environment or habitat was more fully developed in 1895 by the Danish biologist Eugenius Warming, who distinguished plant associations with reference to the water content of the soil, rather than using Merriam's grouping by temperature. Thus "xerophytes" are plants indigenous to arid regions where the water content of the soil is low, while "mesophytes" grow in more moist soils. Of particular importance to estuarine vegetation was Warming's categorization of "hydrophytes," such as cattails or algae, which live immersed in water, and the salt-tolerant "halophytes" like the *Spartina* grasses and eelgrasses of estuaries. Warming was particularly interested in what he called succession—the sequential transformation of landscape by vegetation. [48]

Although succession is a debated concept in modern ecology, terrestrial ecologists around the turn of the century fashioned this idea into an

---

[46] U.S. Department of the Interior, U.S. Geological Survey, "Preliminary Report on the Seacoast Swamps of the Eastern United States," prepared by Nathaniel S. Shaler, in *U.S. Geological Survey Sixth Annual Report*, p. 366.

[47] Nathaniel S. Shaler, *Sea and Land*, p. 246.

[48] Paul B. Sears, *Where There Is Life: An Introduction to Ecology*, pp. 69–73; Henry J. Oosting, *The Study of Plant Communities*, pp. 194, 202, 243, 249; Odum, *Fundamentals of Ecology*, pp. 251–64; Worster, *Nature's Economy*, pp. 198–202.

overarching explanation of nature's economy. To these first-generation ecologists, the tendency of halophytes to "colonize" the bare mud flats of an estuary or lake after a flood's deposition of silt was indicative of a progression toward the eventual development of a high marsh and a later forested borderland. While bogs and mountain lakes were readily recognized as passing from one successional stage to another, sand dunes and salt marshes were also viewed as preliminary and passing steps on the way toward a climax vegetation exhibiting a finer equilibrium with the prevailing climatic regimen. Consequently, turn-of-the-century studies envisioned mangrove swamps as "nature's reclaimers" of land from the sea. This thinking led many investigators astray from Adam Seybert's appreciation for the chemical necessity of marshes in the physical order. Human alteration of tidal marshes could be viewed as an ameliorative process merely speeding up natural succession. In the words of sociologist Lester Ward, man in this sense made the usually wasteful habits of nature more efficient. Far from granting a timeless dignity to the landscape of the tidal marshes, early estuarine studies done under the prevailing ecological concept of succession depicted the coastal wetlands as transient stages in natural development of more orderly and efficient biotic communities.[49]

Shaler's varied studies emphasized the possible conflicts between reclamation for agricultural purposes and dredging for navigation. Drainage of tidal marshes, he acknowledged, "necessarily diminishes the energy and consequent scouring power of the tidal streams." The result could lead to the siltation of the estuary, as Marsh had earlier noted. Nor was Shaler oblivious to the picturesque qualities afforded by marine marshes. "Marshes present a beautiful plain of vegetation," he said. "Nothing . . . is more graceful than the curves of the creeks through which the tidal waters enter and depart." Shaler knew the rich biotic communities that flourished along the shores, "wherever an inlet of the sea is the seat of considerable tidal flow." Conditions there, he explained, "favor the abundant development of animal life."[50]

Despite this apparent appreciation for the oyster, crab, and clam fisheries, Shaler remained an advocate of reclamation, even when calling for an ecological approach to education to combat specialization. In a well-regarded text, he suggested, "In the regions far from the shores . . . life is commonly small in amount and consists mostly of the lower forms." Ex-

[49]O. P. Phillips, "How the Mangrove Tree Adds New Land to Florida," *Journal of Geography* 2 (January, 1903): 10–21; Worster, *American Environmentalism*, pp. 39–53.
[50]Shaler, *Sea and Land*, pp. 238, 243, 250.

hibiting a dismal Malthusian view of the situation, Shaler continued, "In the shallower water near the shores, are the fields to which we may look for help in ages when the world is to be taxed to meet the needs of our kind." Shaler referred to these tidelands as the "debatable ground of the shore zone now occupied by mud flats, marshes and mangrove swamps." Here, he believed, was "a reserve of land awaiting . . . improvement." Shaler estimated that two hundred thousand acres awaited reclamation to provide a more populated country with prime agricultural land.[51]

By the twenties, similar feelings were still being credibly regarded by conservationist Charles Van Hise. He noted that just over 79 million acres of land lay in either swamp or marsh and estimated the cost of reclaiming them at between $4 and $20 per acre. He then suggested four advantages in their reclamation. The market value of the land in rural areas would increase from $20 to $100 per acre, while urban values would rise to $100, or even $500, per acre. Conversion into croplands was a more lucrative use than leaving them as grazing lands or wasted mud. He estimated that food for 10 million people could be raised or that the land could support 50 million residents at the density of Holland. Van Hise's final reason for reclamation was perhaps most persuasive. "By the drainage of wetlands," he noted, "the health of the community will be greatly improved." In this attitude he reflected the national reform sentiment to purify the environment and promote a more controlled cultural advance. "The prevention of disease alone would more than justify the necessary expenditure," he concluded, echoing a sentiment at least as old as Malthus and Rush. This attitude was also reflected in federal policy making.[52]

Frederick H. Newell, as director of the Bureau of Reclamation, became an eloquent spokesman for one facet of the restoration ideal. In 1909, he advocated that "vast areas of swamp and overflowed lands . . . now in private ownership . . . be reclaimed, subdivided, and put in the hands of those men who will cultivate them." Such a plan would improve the community, Newell and his supporters believed. Reclaimed marshes would "be sources of strength to each commonwealth" rather than "breeding places of mosquitoes and other pests, centers of disease and a menace to land values in the neighborhood."[53]

[51] Shaler, *Man and Earth*, p. 93.

[52] Charles Van Hise, *The Conservation of Natural Resources in the United States*, pp. 344–48.

[53] Frederick Newell, "What May Be Accomplished by Reclamation," *Annals of the American Academy of Political and Social Science* 33 (1909): 175.

Like-minded employees of the Department of Agriculture echoed Newell's sentiments. "These waste places," a drainage engineer had suggested, were "rich in fertility and . . . only need draining . . . to make them ideal locations for truck gardening on a large scale." Another official publication warned, "The population of the United States is increasing rapidly and in one State now exceeds 500 to the square mile." In addition to suggesting that the agricultural possibilities were "almost boundless," George M. Warren, a drainage engineer, concluded that "the health, comfort, and well-being of thousands of people must inevitably be promoted." Without pausing to distinguish between tidal marsh and swampland, one enthusiast estimated that there were more than 79 million acres of drainable land in the country. "This gigantic extent of territory," one advocate suggested, was "capable of sustaining a population of fully 10,000,000 people." Engineers of Newell's and Warren's persuasion believed that science and experience had proven the utility of reclamation and that only systematic regional drainage plans were lacking to restore seaside acreage to higher levels of productivity.[54]

Reclamation was widely perceived by Progressive engineers as an act of restoring wasted land to full agricultural potential. Establishing reserves for wildlife propagation was secondary to this promotion of agriculture. Instead of sustainers of fisheries, tidal marshes were viewed as potential truck farms by federal and state bureaus. During the administration of Theodore Roosevelt, conservation of natural resources was characterized by a utilitarian slant. Both reclamation and predator control were considered responsible methods of protecting "desirable" land-use patterns and economically "valuable" species. Growing urbanization, accompanied by industrial and civic pollution of rivers and streams, contributed to conservation's utilitarian emphasis. The larger significance of scientists' and engineers' contributions to federal policies lay in their articulation of a governmental role in convincing the nation's commercial forces to use natural resources wisely.[55]

[54] J. O. Wright, *Reclamation of Tidelands*, 59th Cong., 2d sess., Doc. 820 (Experiment Stations Office), December 3, 1906, p. 374; George M. Warren, *Tidal Marshes and Their Reclamation*, U.S.D.A. Office of Experiment Stations, Bulletin 240, p. 10; Charles Kettleborough, *Drainage and Reclamation of Swamp and Overflowed Lands*, Indiana Bureau of Legislative Information Bulletin No. 2, April, 1914, pp. 10–11.

[55] Samuel P. Hays, *Conservation and the Gospel of Efficiency: The Progressive Conservation Movement, 1890–1920*, pp. 1–26, 199–215, 261–79; Trefethen, *American Crusade for Wildlife*, pp. 124, 129–71; Gifford Pinchot, *The Fight for Conservation*, pp. 43–49, 53–69; Worster, *Nature's Economy*, pp. 262–63; Frederick Newell, "The Engineers' Part in After-the-War Problems," *Scientific Monthly* 8 (March, 1919): 239–46.

Reclamation as a distinct federal and state policy was absorbed during the Roosevelt presidency by the broader federal program of comprehensive riverine management. Conceived by W. J. McGee and Gifford Pinchot, comprehensive riverine management integrated older policies of navigation, flood control, and recreation with newer mandates for reclamation, pollution control, and generation of hydroelectric power. Fishery propagation and wildlife protection were eclipsed in this grand scheme to revive the country's riverine commerce. Despite Roosevelt's creation of the nation's first federal wildlife refuge on Pelican Island, Florida, in 1903, the administration's advocacy of a unified approach to conservation as practiced after the recommendations of the Inland Waterways Commission, in 1907, jeopardized wildlife preservation.[56]

In adding a sophisticated natural-resource plan, these Square Deal advocates of comprehensive riverine management inadvertantly competed with wildlife and fisheries protection for limited congressional appropriations. Concurrently, those opposed to the Square Deal on the local and state levels and within the Army Corps of Engineers desired to keep their lucrative powers under particular, noncomprehensive, rivers and harbors appropriations bills. Caught between these two competing federal coalitions, the older fish and wildlife advocates fell behind riverine managers and the Army Corps in appropriations and power. The isolation of the newly created Bureau of Fisheries under the Department of Commerce from the Biological Survey in the Department of Agriculture was another indicator of the waning influence of traditional conservationists over Progressive conservation policies. In the race for the control of rivers' resources, wildlife-protection advocates failed to hold the nation's attention.[57]

Ironically, the organic approach to government policy making embodied in comprehensive riverine management was the product of scientific advice to federal bureaus. While the fish commission under Baird early contributed the underlying philosophy for the organic and systematic study and propagation of wildlife resources, the influence of ecology waned after Baird's death. The Geological Survey under Powell, the Biological Survey under Merriam, and the Bureau of Reclamation under Newell were major contributors to the organic rejection of laissez-faire resource exploitation.

---

[56] Frank E. Smith, ed., *Conservation in the United States, A Documentary History*, vol. 3, *Land and Water, Part 2: 1900–1970*, pp. 69–111; Hays, *Conservation and the Gospel of Efficiency*, pp. 91–121.

[57] Roderick Nash, ed., *The American Environment*, pp. 37–52.

Initially wildlife protection contributed to this rejection by‘scientists and nature enthusiasts, but an inadequate comprehension of estuaries' biotic functions hampered effective river conservation.

Comprehensive riverine management was a more coordinated and multipurpose way to extract the rivers' numerous resources than was unregulated competitive exploitation. Yet both methods failed to grasp the rivers' economically valuable biotic potential. Fisheries, waste recycling, flood control, and clean water were discounted to enhance the economic viability of narrower developmental schemes. Commerce, dams, reclamation, and agricultural promotion appeared to be more "efficient" uses of the rivers' resources. Because ecological productivity and nutrient recycling were poorly understood by naturalists and ecologists, resource economists lacked a meaningful calculus to fairly measure and equitably divide up the rivers' natural bounty. River degradation continued, with increasingly violent floods, mounting pollution, and persistent declines in fish and wildlife. Reclamation, as part of comprehensive riverine management, increased erosion and sedimentation in estuaries, aggravated saltwater intrusion in coastal areas, and worsened downstream flooding.[58]

Against this remarkable rebuttal of laissez-faire political economy, there would eventually arise a new awareness of the limits of federal power and the natural confines within which social engineering and technological change may safely occur. Without this successful political challenge to the myths of rugged individualism, democratic self-reliance, and limited government, though, the national awareness of ecological realities that emerged after World War Two could not have created the political consensus that temporarily preserved the coastal wetlands from further reclamation. Somewhat ambivalently, tidal marshes had been recognized as essential habitats for fish and fowl while still being viewed as wastelands—after 1890, they were seen as a surrogate frontier of reclaimable farmland and homesteads to fulfill the popular promises of Progressive political and social ideology.

[58] Brewer, "Brief History of Ecology," pp. 9, 11–13; Smith, *Conservation*, III, 75–110; Aldo Leopold, *A Sand County Almanac*, pp. 202–22; Nash, *American Environment*, pp. 97–98, 131–51; Darrah, *Powell of the Colorado*, pp. 301–303; Trefethen, *American Crusade for Wildlife*, pp. 123–25; Mel Scott, *American City Planning*, pp. 17–37.

6

The New Ecology and a New Ecological Ethic

Some Engineers are beginning to have a feeling in their bones that
the meanderings of a creek not only improve the landscape but are a
necessary part of the hydrologic functioning. The ecologist sees
clearly that for similar reasons we can get along with less channel
improvement on Round River.

—Aldo Leopold, 1949

AT the turn of the century ecology and economics agreed on the classification of tidal marshes as obstructing wastelands. Resource economics determined the "best" use of land solely from human and utilitarian perspectives, and initial findings of terrestrial and marine biologists supported policies to make human "improvements" over nature's perceived inefficiency. However, in the 1930s and 1940s a number of scientists, including wildlife biologist Aldo Leopold, would begin to demonstrate that the political and economic values assigned to coastal wetlands and other wild areas conflicted with their biological integrity. By the late 1960s the nation's marshes would have attained the dignity first accorded them by Adam Seybert in 1798; legal protection for coastal wetlands would be based on findings of a "new ecology" that reflected significant scientific discoveries in physics, chemistry, geology, and biology.

The values of the new ecology, though, faced formidable opposition from the Progressive consensus on human-centered, utilitarian criteria. Progressives considered grazing on old salt meadows, for example, a less efficient use of tidal marshes than reclaiming and draining those ranges for truck farming. Near urban areas, truck farms were held to be a less profitable use of developable lands than resorts, residences, or industrial sites. Reclamationists felt that tidal marshes were inefficient natural obstacles to municipal expansion, a public trust for breeding mosquitoes and diseases, submerged real estate ready for speculative transformation into suburbs, seaside resorts, or public ports. Especially in government, fiscal conservatism fostered the search for useful applications of scientific research to improve human economy. Thus the uses of tidal marshes for public parks took precedence over the preservation of wetlands for wildlife refuges, and, particularly relative to the suburban development advocated by Progressive engineers, wild seashores and especially tidal marshes became expendable.[1]

The adverse ecological effects from suburban expansion were caused largely by three ingredients of the Progressive desire for a more equitable society: independent home ownership, the notion of garden cities as the marriage of city and country, and communal financial responsibility for enhancing impoverished lands. Progressives replaced the older pastoral ideal of family farms stretching from sea to sea with a novel idealization of the suburbs as the "middle landscape" between the cluttered decay of urban environs and the depressing isolation of the rural countryside. Neither the cities teeming with recent migrants nor the farms from which they fled were the preferred residences of modern Americans.[2]

When setting the nation's postwar agenda in 1919, Frederick Newell cited "an adequate land policy" and "sanitary and comfortable housing" as two of the ten goals reflecting Progressive ideals. These aims included what was considered the best of the country and the best of the city and were based on an unshakable faith in the ability of science to understand and control the physical world. Harnessing hydraulic energy for human social advancement reflects best this pragmatic faith in positive science. Progressive river conservation required dams to provide the needed hydroelectricity to light safer cities, warm homes with less dirt, and produce the increasing number of labor-saving devices to meet the growing domestic demand. Public ownership and control of the vast hydroelectric resources of the public domain became the New Deal's leading policy for ensuring the nation's future industrial growth.[3] This policy, called comprehensive riverine management, was designed to balance the competing uses of watersheds for recreation, wildlife protection, flood control, hydroelectric power generation, dredging, and reclamation.

In the quest for greenbelt communities, seaside resorts, and hydro-

---

[1] Charles Van Hise, *The Conservation of Natural Resources in the United States*, pp. 183, 344–48; Nathaniel Southgate Shaler, *Man and the Earth*, pp. 87–100, 167–69.

[2] Leo Marx, *The Machine in the Garden*.

[3] Frederick Newell, "The Engineers' Part in After-the-War Problems," *Scientific Monthly* 8 (March, 1919): 239–46.

electric power, the economic demands of the nation would eventually confront the ecological needs of our many estuaries.[4] Both the dreams of garden suburbs and the utilization of the nation's long-neglected waterways would collide with the biogeochemical functions of estuaries—the long-neglected aspect of the river's physical relation to the sea.

In the meantime, developments on a different front furthered the breakdown of the consensus on the value and legal status of lands along the coasts. During the depression and war years coastal wetlands and their adjacent submerged lands were brought to the national attention by government action to promote oil exploration. Coastal areas became the focal points for a social confrontation lasting more than three decades, involving all three levels of government, and engendering a cultural redefinition of terms like equity, property, community, and responsibility. From 1937, when federal authorities claimed an overriding national interest in the submerged lands lying along the coast, until the summer of 1968, the consensus reserving tidelands to the states as a public trust fractured.

Upon the announcement of Interior Secretary Harold Ickes's decision to grant offshore oil leases to petroleum corporations, coastal states launched a counteroffensive in the courts. Having lost their appeals in the Supreme Court in 1947 and wary of President Truman's proclamation declaring the Continental Shelf federal property in 1945, the states recovered their rights to the foreshore through congressional legislation in 1953. Although the participants have called this argument over state control of the Continental Shelf the tidelands controversy, the ownership of the actual intertidal zones was never in question. Rather, permanently submerged lands out to the three-mile limit (or ten-mile limit, off Texas) embroiled local and national interests. The very problem of reckoning the tide lines along the shore from which to measure state lands confounded the implementation of the Submerged Lands Act of 1953, which quitclaimed the sublittoral zone out to the three- or ten-mile limit to the states.[5]

The initial clash between officials in Washington, D.C., and the states of Louisiana, Texas, and California had involved the submerged "tide-

[4]Mel Scott, *American City Planning*, pp. 89–90, 176, 206–10, 215–20, 337–42, 436–37, 457, 580–86.
[5]Ernest R. Bartley, *The Tidelands Oil Controversy*, pp. 213–46; Francois Uzes, *Chaining the Land: A History of Surveying in California*, pp. 131–33; Aaron L. Shalowitz, *Shore and Sea Boundaries*, I, 3–20, 31–63, 105–12, 115–73, 187, 362–67.

lands" because of offshore oil drilling, but the final federal-state confrontation involved the intertidal zones and their fringing wild tidal marshes. The inexorable national demographic shift to the coastal zone between 1940 and 1970 made the political disposition of estuarine resources an important issue. Consequently, the 1953 reassertion of states' rights over the tidelands was not the final resolution hoped for by its supporters. While state land ownership within the estuaries had been reaffirmed and extended from the high-tide line to the three-mile limit, water use became the driving wedge reopening the question of governmental responsibility for coastal-zone resources.[6]

Federal authority had grown to include national planning under the policy of comprehensive riverine management. That policy, sustained by the Supreme Court in 1936, required the maintenance of water quality standards, as part of a multiple-use ideology. Growing demand for clean water in the coastal zone was the seam along which governmental cooperation tore and reopened the issue of specific intergovernmental responsibilities within coastal areas and river basins.[7] Water-related issues were thus numerous and entwined on the local, state, and federal levels. The decline in fisheries involved conservation, deteriorating water quality required regional planning, dredging involved the army, and local governments encouraged the filling of marshes.

Concurrent conceptual changes in economics also encouraged federal manipulation of fiscal and monetary policy to underwrite a continuously expanding industrial and technologial system. These policies particularly promoted the coastal expansion of the nation—its suburbs, highways, airports, recreational facilities, marinas, and energy production facilities. Economic expansion, increasing per capita demands for land and water, additional services, and synthetic technological innovations all rested on

[6] Samuel P. Hays, *Conservation and the Gospel of Efficiency: The Progressive Conservation Movement, 1890–1920*, pp. 2–4, 114–21, 262–71; *Ashwander* v. *Tennessee Valley Authority*, 297 U.S. 288, cited in Alfred H. Kelly and Winfred A. Harbison, *The American Constitution: Its Origins and Development*, pp. 747, 789; Roderick Nash, ed., *The American Environment*, pp. 131–51, 178–83; John H. Davis, "Influences of Man upon Coast Lines," in *Man's Role in Changing the Face of the Earth*, ed. William L. Thomas, Jr., pp. 504–21.

[7] Wesley Marx, *The Frail Ocean*, pp. 9–23, 59–109, 143–69; Institute of Ecology, *Man in the Living Environment: A Report on Global Ecological Problems*, pp. 34–37, 219–59; Donald E. Carr, *Death of the Sweet Waters*, pp. 43–55, 66–110; John Teal and Mildred Teal, *Life and Death of the Salt Marsh*, pp. 199–262.

federal aid and assumed the desirability of growth. After 1945, this growth required a further reliance on old energy sources and the development of spin-offs from nuclear armament research.[8]

Given all these demands, coordinated federal policies eluded planners, and the jurisdictional dispute between coastal states and the national government over offshore oil exploration delayed a consensus on coastal development for twenty years.

Furthermore, by the late 1940s the unrestrained pursuit of the Progressives' ideals was sharply criticized by wildlife biologists and urban planners concerned over water quality, adequate marshes for spawning habitats, and continued urban sprawl. People like Aldo Leopold interpreted these ideas and ecological values into a widely understood plea for preservation of all wild things and fostered a growing ecological revolt against prevailing cultural ideals. The Progressive dream was challenged, ironically, by the very suburban residents whose communities stood as testaments to the tenacity of Progressive values.[9]

Ecological studies eventually concluded that estuaries and their bordering salt marshes are some of the most productive biological communities on earth. This change centered on the role of water as a habitat for life and the use of energy as a measure of natural success in converting sunlight and inorganic nutrients into life's building blocks. The story is one of several accretions in scientific investigations ranging from biochemistry and biophysics to marine ecology and microbiology. Heavily influenced, too, by economics and physics, ecology emerged after World War II with a rigorous critique of technology resting on a new appreciation for the roles of recycling and waste in natural systems.[10]

The concepts of energy transfer and waste recycling in the new ecology replaced the older biogeological community-succession notion that suggested marshlands were only transitional stages of vegetational change.

[8]Herbert Stein, *The Fiscal Revolution in America: 1930–1968*; Barry Commoner, *The Closing Circle: Nature, Man and Technology*, pp. 45–62, 122–75; Carroll W. Pursell, Jr., *Readings in Technology and American Life*, pp. 423–59; Seymour Melman, *War Economy of United States* (New York: St. Martin's Press, 1971).

[9]U.S. Congress, House, Committee on Merchant Marine and Fisheries, *Estuarine and Wetlands Legislation. Hearings Before the Subcommittee on Fisheries and Wildlife Conservation*, 89th Cong., 2d sess., June 16, 22–23, 1966, pp. 1–127.

[10]Donald Fleming, "The Roots of the New Conservation Movement," *Perspectives in American History* 6 (1972): 64–81; Donald Worster, *Nature's Economy: The Roots of Ecology*, pp. 256–348; Eugene P. Odum, "The New Ecology," *Bioscience* 16 (July, 1964): 14–16.

The full force of this conversion was embodied in A. G. Tansley's concept of an ecosystem. This Oxford biologist, writing in 1934, hoped to replace the older organismic concept of community (biocoenosis) with a more quantitative idea amenable to systematic scrutiny, measurement, and prediction, along the line of quantum physics research.[11]

Biological communities as they interact with geographical habitats were described by the term *ecosystem*. As the focus of study, the ecosystem was a theoretical description of both the dead and dying (organic) and never-living (inorganic) portions of any environment. An ecosystem could comprise a small patch of salt marsh, an entire estuary, a rain forest, a coral reef, or even the earth. This nonspecificity allowed comparisons of the functional aspects of any living community impossible for the earlier descriptive naturalists and ecologists, who had catalogued and categorized habitats and species. Not only could the diversity and numbers of species of different ecosystems be compared, but now the behavior, productivity, and even efficiency of one system was comparable with that of another—for instance, forests with estuaries. Freed from the counting of individuals and niches within communities, scientists could now concentrate on the organizational levels among the biological communities of a given ecosystem fostered by different physiological and geological factors. Finally, Tansley's approach encouraged the search for a common denominator of exchange representing the relation of the inorganic to organic worlds. The older food web of naturalists gave way to energy exchange as the guiding image in ecosystem organization and change.[12]

Ecosystem logic, by inverting Einstein's energy-mass equivalency, traced the energy photosynthetic organisms used in breaking the hydrogen bonding of the water molecule and thereby determined the yield of photosynthesis in terms of mass, in this case biomass. When they measured the biotic community's efficiency at converting radiant energy into food and fiber, ecologists realized that marshland communities utilized solar energy more efficiently than other communities and distributed it so effectively that wildlife flourished. Ecological studies tried to measure and predict the rate of solar energy transformation in photosynthesis and food's subsequent distribution throughout the system. Quantum mechanics had re-

[11] Worster, *Nature's Economy*, pp. 294–305; Eugene P. Odum, *Fundamentals of Ecology*, pp. 34–85; Thomas Odum, *Energy, Power, and Society*, pp. 104–37, 174–205.

[12] Warder C. Allee et al., *Principles of Animal Ecology*, pp. 13–24; Fleming, "Roots of the New Conservation," pp. 20–24; Odum, *Fundamentals of Ecology*, pp. 8–33.

vealed that light was measurable in discrete packets, and therefore the amounts and wavelengths necessary to trigger photosynthesis could eventually be predicted.

The laws of thermodynamics as applied to energy conversions—from sunlight to food and at every level of animal consumption along the complex web of feeding relations leading to top-level predators—meant that the total energy bound up in the prey could never be made fully available to the consuming organism. Some energy was always needed to sustain the life processes of respiration, while at every transformation of energy from one form to another heat was lost. Heat, the least usable form of energy, cannot be economically converted into a more efficient form. Consumption patterns within ecosystems were referred to as trophic levels, and none proved more elusive yet essential than the decomposers recycling molecular nutrients essential for plant and animal maturation. The least understood lesson of the new ecology was the necessary role viruses, microbes, bacteria, and fungi play in creating a successful, life-sustaining dynamic equilibrium.[13]

Ecological research thus experienced a threefold shift of emphasis beyond the single-faceted biocentric revolution of the nineteenth century. The changes from descriptive to functional explanations of nature, from the study of individuals in communities to levels of organization in ecosystems, and from food webs to energy flow all suggested a larger than biological view of life. This intellectual movement represented an integration of the older biocentric outlook with the quantum physics of Albert Einstein and Niels Bohr, augmented by microbiological and oceanographic research after World War II.[14]

The ecological revolt, the product of an ecocentric view of humanity's relation to nature, embodied all of the basic tenets of twentieth-century science, coherently grafted to the post-Darwinian conceptions of the biological order. The practical implication of this intellectual synthesis of diverse scientific perspectives may be appreciated better by looking at a small lake than at a highly complex river mouth. The lake as such a small unit of ecological study was described by Stephen A. Forbes, the head of the Illinois State Natural History Survey, as early as 1887. His paper, "The

---

[13] Eugene Odum, "New Ecology," pp. 14–16; Selman Waksman, "The Role of Bacteria in the Cycle of Life in the Sea," *Scientific Monthly* 38 (1934): 35–49; Fleming, "Roots of the New Conservation," pp. 34–39.

[14] Odum, "New Ecology," pp. 14–16; Worster, *Nature's Economy*, pp. 185, 302.

Lake as Microcosm," discussed the feeding relationships existing between species in a tiny Illinois lake and suggested that the lake "forms a little world within itself" where the earth's "elemental forces are at work." Although his study reflected the biocentric ideal (and the Darwinist emphasis of his times), Forbes pioneered the use of a discrete unit of nature to explain the population balances maintained by predator-prey relationships. He thus established the usefulness of studying an integrated geographic unit. Later ecologists used Forbes's method to understand the changing reciprocal relationships of life, landscape, water, and air.[15]

Forbes's pioneering path was followed and enlarged upon by Raymond Lindeman, at Yale, in 1942. Lindeman entitled his study of a cedar bog lake in Minnesota "The Trophic-Dynamic Aspects of Ecology." Before he died at the age of twenty-seven, Lindeman, as one of the brightest minds in the younger generation of ecologists, had done work that was seminally important to the development of the new ecology. Having grouped all the resident organisms into trophic levels, Lindeman measured the amount of energy loss at each level of transference in order to arrive at the measure of caloric energy needed to sustain the accumulated biomass at each level. Because plants capture the radiant energy of the sun and transform it into living structures or biomass, they are the producing foundation of ecosystem organization. Those animals existing solely on plant material, or herbivores, are classed as primary consumers. Thus, phytoplankton and cattails are considered producers of the lake environment, while copepods and insects feeding on the plants are classed as primary consumers. Fish and birds prey on the copepods or insects of the lake and thus are classed as secondary consumers. Those animals preying upon the fish, such as bears, or upon the birds, such as wild cats, are classed as tertiary consumers, and so on up this overly simple energy pyramid.[16]

Lindeman discovered through quantified analysis that the decomposing unit of the ecosystem is a significant retriever of unused energy from all levels of the trophic structure. What scientists had always called detritus, or the decayed remains of plants and animals, was a rich source of nutrients and food in aquatic ecosystems. Thus, any animals that survive by eating the decomposing matter of more complex life forms have avail-

[15] Stephen A. Forbes, "The Lake as Microcosm," *Illinois Natural History Survey Bulletin* 16 (1925 reprint of 1887 ed.): 538–50.
[16] Raymond Lindeman, "The Trophic Dynamic Aspect of Ecology," *Ecology* 23 (1942): 399–418; Worster, *Nature's Economy*, pp. 306–11.

able a considerable energy supply. Indeed, detritus feeders living on terrestrial, aquatic, or marine humus are important parts of the productivity
of lakes, bogs, or marshes. This study confirmed earlier marine bacterial
work by Selman Waksman (1934) suggesting that the muds of the coastal
zone are rich in organisms that break down complex structures, making the
necessary nitrogen, phosphorus, and carbon available to the next generation of plants and animals.[17]

Later studies confirmed important ecological functions working in
every ecosystem with varying degrees of efficiency. These efficiencies
vary according to the timing of population increase, the differing structures
of individual species, and the stresses endured from the surroundings. But
every ecosystem adjusts to the reality that energy degrades at every level of
transformation, even what comparatively little energy is utilized by each
trophic level from the one below.

Energy dissipation in living systems is called the law of the ecological
tithe and is the biotic equivalent of the second law of thermodynamics in
physics. In every ecosystem, the most efficient transformation of energy
occurs at the level of primary productivity, when plants convert sunlight
into life. This primary productivity was shown as sharply increasing at the
earliest stages of ecosystem development, then tapering off as the lake
aged, its oxygen was depleted, and sedimentation or evaporation hastened
its transformation into a peat bog. The energy, however, would remain
stored in the decayed matter of the bog, available to humanity in the form
of peat moss for fuel.[18]

In the same year that Lindeman's study appeared, Harold Sverdrup,
director of the Scripps Institution of Oceanography, codified the preceding
century's findings in oceanography. His description of the physical properties of seawater suggested the levels of stress to which marine life must
adapt. Temperature, pressure (increasing with depth), and salinity were
the three broad measurements defining the various marine ecosystems. He
quantified work in the tradition of Maury and Agassiz, facilitating the
cross-disciplinary investigations upon which oceanography was founded
and thrived.

Sverdrup described the oceans primarily as a vast reservoir for the
storage of solar radiation. He referred to this as the oceans' high specific
heat because seas warmed and cooled more slowly than their adjacent

[17] Waksman, "The Role of Bacteria," pp. 35–40.
[18] Odum, *Fundamentals of Ecology*, pp. 9–10, 24–33, 359–60, 484–97.

lands. This differential gain and loss of energy by land and sea is the vast engine driving the freshwater cycle. The single most important circuit of energy on earth, sustaining life and creating climates, this cycle accounts for the evaporation of seawater into fresh water. Saline waters are heavier and denser than the fresher flows of upland rivers or marshes. The brackish mixing of these two kinds of water in estuaries posed distinct evolutionary problems for the earliest marine organisms, which eventually colonized rivers and shores. Scientists began to study the biota of estuaries to understand more fully the anatomical changes required by species that must tolerate fluctuating stresses from hypersaline tides or upriver floods. Slowly, ignorance of estuarine dynamics yielded to more exacting scientific research.[19]

Gordon Gunter's studies of the Gulf Coast plankton fisheries revealed the twin effects of salinity and pollution on life cycles. His findings demonstrated that shrimp migrate from the offshore waters above the Continental Shelf to near-shore estuarine waters during their life cycle, subjecting the juveniles to stress from temperature, pressure, and salinity. In the brackish waters of estuaries, young or adolescent shrimp mature, feeding and falling prey to fish as they are carried up the estuary by incoming tidal currents. Gunter's studies demonstrated that silting up of channels and the accumulation of human waste in particularly sluggish river mouths contribute to water conditions that foster the growth of toxic single-celled dinoflagellates. Gunter studied a collapse of the shrimp, crab, and oyster fisheries in Galveston Bay in 1939 that resulted from the depletion of oxygen by a tide of these single-celled creatures. Both migratory birds and fish are seasonally attracted to the estuaries and adjacent marshes of the coastline, adding to the diversity, productivity, and utilization of the energy resources in these regions.[20]

During the 1940s the groundwork was established for an eventually rigorous and sustained critique of one-dimensional planning and the human-centered philosophy of science around which governmental policies were designed. These criticisms came during an unprecedented period of single-purpose technological development, which yielded antibiotics,

[19] Harold Sverdrup, *Oceanography for Meteorologists* (New York: Prentice-Hall, 1942), pp. 1–35; A. S. Pearse, *The Emigrations of Animals from the Sea*, pp. 3–7, 17, 22–37.

[20] Gordon Gunter, "Some Relations of Faunal Distributions to Salinity in Estuarine Waters," *Ecology* 37 (July, 1956): 616–19; Maurice Burton, *Margins of the Sea* (New York: Harper Brothers, 1954), 4–6.

DDT, synthetics, and the atomic bomb. The perspective fostered by the ecocentric revolution depicted these miracle technologies as disturbers, rather than facilitators, of biological health.

Then, as scientific knowledge about the reciprocal ties of humans to the physical world accumulated, a wildlife biologist with an extraordinary ability to interpret these findings promoted a new ecological ethic. Aldo Leopold, understanding how energy, land, and water form the exploitable milieu within which all life flourishes, stressed the sacred cohesiveness of the earth organism. Because each individual's survival depends on the functioning biotic integrity of physical and chemical cycles, Leopold conceived of the earth with its living endowments and infinite landscape variations as a single being that affords all other creatures nourishment.

Throughout the earth's history the coevolution of plant, animal, and human communities has been sustained by integrating geophysical and biochemical forces; these protean ecological cycles eventually assured successful human domination worldwide. Increasing human ecological interference, Leopold argued, has made it imperative that social ethics evolve and create new caretaker institutions for the earth's survival. The older adversarial, utilitarian, and aesthetic relationships with nature must be superseded by earth-sustaining, ecologically reasonable ethics. In his own mind, Leopold was merely repeating the ancient prophetic wisdom of Ezekiel and Isaiah, who had warned against the destruction of the land. But Leopold's land ethic also incorporated the new scientific insights to quicken the formulation of a new estuarine preservation ideal.[21]

Leopold, like other scientists of the day, expressed a growing disillusionment with Progressive resource ideology and its consequent destruction of wetlands. The lake that Stephen Forbes had written of in 1887 was drained by 1925. The agricultural survey incident to the census of 1920 revealed that 10 percent of the nation's 79 million acres of drainable land were classified as tidal marshes. Especially after 1940, estuaries lay in the region of greatest population growth and density increase.[22] In the face of this concentration and destruction of America's wetlands and as an alternative to Progressive utilitarianism, Leopold advocated a land ethic under which individual landowners would take personal moral responsibility for the land's health.

[21] Aldo Leopold, *A Sand County Almanac: With Essays on Conservation from Round River*, pp. 237–39.
[22] Teal and Teal, *Life and Death of the Salt Marsh*, pp. 239–40.

His criticism of Progressive policies was forceful and direct. By assessing immediate economic benefits only, they tended to "ignore, and thus eventually eliminate, many elements in the land community that lack commercial value, but that are . . . essential to its healthy functioning." They were shortsighted and failed to calculate unintended side effects. Finally, they elevated man to a position above, and strangely isolated from, the environment of which he was actually, Leopold insisted, an integral part.[23]

On the first point, he argued that "a system of conservation based solely on economic self-interest is hopelessly lopsided." The gulf between economic values and ecological necessities was a major theme in Leopold's writings. "Lack of economic value is sometimes a character not only of species or groups," he wrote, "but of entire biotic communities: marshes, bogs, dunes, and 'deserts' are examples." Since these lands lacked economic value, especially in comparison with urban areas, "we have relegated some of them [the communities] to ultimate extinction over large areas."[24]

Furthermore, by adopting narrow calculations of economic utility, Progressive resource economists arrived at improper cost-benefit analyses. In fact, they often classed as useless the essential elements of a functioning biophysical equilibrium, and nowhere was the contrast of values between ecology and economics more apparent than in policies concerning estuarine tidal marshes or interior wetlands. For example, as Leopold and other ecologists recognized, mud and decomposed matter were necessary to the maintenance of healthy biotic communities. *Waste* was a value-laden term for matter not readily used within the human economic system—but recycled for use in nature's economy.

During the late 1920s the drainage of swamps and overflowed lands to create farmland had begun to be challenged on economic grounds alone. Why should the population underwrite the conversion of more costly acreage into farmland when the country had been suffering from overproduction of crops and subsequent farm depression since 1921? Leopold expanded the critique of wetlands reclamation policies by pointing out a number of unintended side effects. Crops grown on indiscriminately reclaimed marshlands were poor, and "expensive ditches added an aftermath

---

[23] Leopold, *Sand County Almanac*, p. 251.
[24] Ibid., pp. 249, 251.

of debt" that eventually drove the farmers out. Drained lands were more vulnerable to disaster: "Peat beds dried, shrank, caught fire." When a series of dry years set in, Leopold reminded, "not even the winter snows could extinguish the smoldering marsh."[25]

Moreover, the removal of wetlands had three drastic hydrologic consequences. First, as settling areas for water in seasonal flood-control reservoirs, swamps facilitated the percolation of groundwater into underground aquifers. Many municipalities relied on underground water sources and would suffer if these supplies were not replenished. Second, once the land along flood plains was drained, the soil had a tendency to shrink and subside. This subsidence placed the reclaimed areas in greater dangers of flood than previously, and this required the construction of new levees or the elevation of existing ones. Finally, areas downstream from drained swamplands were also made more vulnerable to flooding since the heavier and more rapid flow of water washed away streamside plants that had maintained riverbanks against erosion. "The creek is now able to carry off more flood water, but in the process we have lost our old willows. . . ." Leopold wrote. And with the willows, more than the trees themselves disappeared; more than the banks they protected from erosion were gone. A complex habitat for birds and other wildlife also passed out of existence.[26]

Leopold believed that part of government's miscalculation of the utility of environmental alterations came from its shortsightedness in not recognizing that removal of native vegetation did more than merely disrupt the prolifically diverse wild patterns of energy transfer. Introduction of agriculture into former marshlands represented what ecologists called a "disclimax"—that is, the maturing natural vegetation was replaced by grains and other non-native plants intended for harvest or pasturage. Leopold saw the initial vegetational disruption of the landscape as relatively innocuous, an "Arcadian age for marsh dwellers," during which "man and beast, plant and soil lived on and with each other in mutual toleration, to the mutual benefit of all." Indiscriminate and large-scale reclamation projects, however, were another matter; Leopold characterized the programs of such conservationists as Nathaniel Shaler, Frederick Newell, and Charles Van Hise as "an epidemic of ditch-digging and land-booming." The culturally induced vegetational changes resulting from their programs could even-

<hr>

[25] Ibid., p. 106.

[26] Aldo Leopold, *Round River: From the Journals of Aldo Leopold*, ed. Luna B. Leopold, p. 165.

tually diminish not only natural biotic diversity but also human psychic health, robbing future generations even of the meaning that wilder areas give to cultivated landscapes. Solitude, not easily obtained in civilization, is invaluable, he insisted, but its importance "is so far recognized . . . only by ornithologists and cranes." [27]

Indeed, to Leopold, the psychic value of wilderness went beyond mental health to an almost spiritual dimension. He judged the quality of a landscape by the number and variety of its wildlife, and particularly wildlife that avoided human company, such as cranes, clapper rails, sea otters, or seals. "The quality of cranes," he wrote, "lies . . . in this higher gamut, as yet beyond the reach of words." Because the crane is a living remnant of the Eocene Epoch, "he is the symbol of our untamable past, of that incredible sweep of millennia which underlies and conditions the daily affairs of birds and men." A crane marsh consequently "holds a paleontological patent of nobility." From an ecological point of view, Leopold was stressing the value of diversity, not just for the birds or the landscape, but for the peace of mind and continued evolutionary health of mankind. He concluded, "The ultimate value in these marshes is wildness, and the crane is wildness incarnate." [28]

Just as Progressive conservationists failed to appreciate the importance of the natural world for the quality of human life, so they failed to see how human beings were integrated into the rest of nature. Instead, they treated wildlife and wild landscapes alike as adversaries of civilization. For too many years, Leopold insisted, the bureaucratic approach to the political economy of natural-resource utilization and allocation had engendered a feeling of contest—man versus the environment—and even the language chosen reflected this adversarial stance and the subordination of other forms of life to the convenience of man. Pest control required the use of "pesticides." Weed killers were called herbicides, as if an unending war had to be waged against plants that are a normal biotic response to human disturbance.

Leopold believed that American intellectual history, education, and bureaucratic practices concerning the environment had produced a generation of self-centered individuals who misunderstood how the demands of society and economics competed with and unraveled the landscape's biotic

[27] Leopold, *Sand County Almanac*, pp. 106, 107.
[28] Ibid., pp. 102, 103, 107–108.

integrity. Modern urban and suburban dwellers had "outgrown" their ties to the land, and their autistic attitudes were reflected in many ways, from advocacy of flood control and reclamation to the predator control programs of state and federal wildlife management. Leopold noted that "we assume [waters] have no function except to turn turbines, float barges, and carry off sewage." To the contrary, he pointed out, "waters, like soil, are part of the energy circuit. Industry, by polluting waters or obstructing them with dams, may exclude the plants and animals necessary to keep energy in circulation." [29]

Using the settlement of the Colorado River mouth's innumerable islets as an example, Leopold commented on the consequences of the seemingly endless war of government agencies, ranchers, and farmers against large predators. "By this time the Delta has probably been made safe for cows," he wrote. "Freedom from fear has arrived, but a glory has departed from the green lagoons." To him, thriving animal life symbolized the health and integrity of the ecosystem: "I cannot recall feeling, in settled country, a like sensitivity to the mood of the land." It was crucial, he felt, that a land ethic be developed that would change "the role of *Homo sapiens* from conqueror of the land-community to plain member and citizen of it." [30]

All of these shortcomings in Progressive conservation policy stemmed from an underlying failure of vision. Policy makers had not appreciated what scientists were revealing to be a thorough interdependence and interrelatedness of the elements of life on earth. For Leopold and others, Progressives' "improvements" were short-circuits, not sustainers, of the bioenergetic integrity of ecosystems.

To illustrate this organic interconnectedness Leopold chose the parable of "Round River" from the legends of Paul Bunyan. Round River, he explained, was a "river that flowed into itself . . . in a never-ending circuit." Its current "is the stream of energy which flows out of the soil. . . ." [31] The land itself "is not merely soil; it is a fountain of energy flowing through a circuit of soils, plants, and animals." [32] In Round River's circuit of life, "a rock decays and forms soil; in the soil grows an oak, which bears an acorn, which feeds a squirrel, which feeds an Indian."

[29] Ibid., pp. 240, 255.
[30] Ibid., pp. 152, 155, 240.
[31] Ibid., p. 188; Leopold, *Round River*, p. 158.
[32] Leopold, *Sand County Almanac*, p. 253.

Upon the man's death, his nutrients return to the soil "to grow another oak." The Round River ecological metaphor thus portrays continuing human dependence on the earth as a symbiotically coevolving relationship that energetically and genetically ties existing biological communities to their biotic ancestors and abiotic predecessors. The circuit of nature's economy is "like a slowly augmented revolving fund of life." Round River "grows ever wider, deeper, and longer."[33]

Earlier ecologists, including those under whom Leopold had studied at the Yale School of Forestry, had called "this sequence of stages in the transmission of energy a food chain," but, Leopold argued, "it can be more accurately envisioned as a pipeline." Energy "leaks" from this pipeline, he explained, and "only part of the energy in any local biota reaches its terminus." Some energy is sidetracked, as when a "squirrel drops a crumb of his acorn which feeds a quail, which feeds a horned owl, which feeds a parasite."[34]

Behind Leopold's deceptively clear natural descriptions were several assumptions that reflected the contributions of physics, microbiology, and economics to the study of ecology. Physicists through thermodynamics measured the amount of energy lost in each conversion, while microbiologists explained decomposition's vital role in making nutrients available to what the economists called nature's "producers," or plants, and "consumers," or animals. Like his colleagues, Leopold attempted to move beyond the food-chain conception of biotic relations, but his contribution was to depict the energy flow from mud to fish to fowl in less abstract and technical language than the trophic structure of the new ecology. Nonetheless, both the abstract terminology of the scientist and the descriptive language of the naturalist reflected the same underlying realities.

The physical conditions necessary for life to flourish were "chains of dependency" in Leopold's terminology—"a maze of services and competitions, of piracies and cooperations." Over time these interconnections became more complex, and entire biotic communities changed by a process called evolution. The stability of a given biotic community over time depended on the successful coevolution of its producing, consuming, and decomposing organisms and the cohesion of their energy transfers. "This maze is complex," explained Leopold; "no efficiency engineer could blue-

---

[33] Leopold, *Round River*, pp. 159, 162; Leopold, *Sand County Almanac*, p. 253.
[34] Leopold, *Round River*, p. 160.

print the biotic organization of a single acre." Leopold recognized the difficulty of the task of ecologists in attempting "to convert our collective knowledge of biotic materials into a collective wisdom of biotic navigation." Yet doing so, understanding biotic communities as energy systems, was the central problem for resource economics, conservation engineering, and fish and wildlife protection. He concluded, biotic navigation "is conservation."[35]

To Leopold, a landscape possessed biological integrity if the large predators, such as the jaguar or coyote, were sustained by the uninterrupted flow of energy from their prey. Like the dolphins or harbor seals of an estuary, the great cats or bears represented the pinnacle of the food web that transferred energy and nutrients from one animal to another. All these energy pyramids were made possible by the transformation of radiant energy into chemical energy through the photosynthetic action of phytoplankton, algae, or plants and the recovery of essential elements by bacteria. But Leopold used terms more readily understandable than levels of ecological organization or efficiencies of energy transfer to sum up the need for marshes and other wild communities: they should be islands of diversity and wildness in a sea of uniformity and civilization.

In a parable entitled "Marshland Elegy," Leopold pondered human capacity to comprehend and nurture the landscape's essential character. "Our ability to perceive quality in nature," as in art, must be cultivated, he mused. It "expands through successive stages of the beautiful to values yet uncaptured by language." The appreciation of certain life forms was beyond the grasp of uninformed minds preoccupied with the one-dimensional management of their own lives or occupations. The odious consequences of such a lack of popular regard for wild things made the task of conservation extremely difficult, particularly in a democracy, where popular attitudes were political capital.[36]

Leopold argued that education held the key to changing the autistic bent of influential policy makers. As an educator himself, Leopold spent much of his life trying to introduce the view of biotic communities as a stream into the educational system. He defined ecology as the "science of relationships" and insisted that it called for "a reversal of specialization." He realized that such an interpretation cut across modern curriculum's divided disciplines and that therefore the task for educators was formidable.

---

[35] Ibid., pp. 159, 162; Leopold, *Sand County Almanac*, p. 189.
[36] Leopold, *Sand County Almanac*, p. 102.

But it was necessary. "Does the educated citizen know he is only a cog in an ecological mechanism?" he asked. The purpose of environmental education was to prepare people to live within and act responsibly toward their particular biotic community. This preparation would give the individual a choice: "If he will work with that [ecological] mechanism his mental wealth and his material wealth can expand indefinitely." Otherwise, Leopold warned, "it will ultimately grind him to dust." [37]

To Leopold, the development of individual consciousness was essential, since government conservationists did not possess a land ethic. He was skeptical of the ability of state and federal agencies or of private industry to adopt and implement faithfully ethical values for protecting the land's biotic integrity. All such organizations were "too busy with new tinkerings to think of the end effects." [38]

Leopold criticized the government, even in its role as conservationist, for one-dimensional planning. He challenged the secondary position to which Progressive planners had relegated the fish and bird protectors in bureaus that predated the inception of "conservation," and he expressed disenchantment with state and federal policies to restore worn-out lands. McGee's and Pinchot's multipurpose planning proposal for regional watershed management notwithstanding, one-dimensional planning had remained the norm in government policy making, partially because of specialization within federal or state bureaus and in congressional committees. [39]

The single notable exception to such myopia in government planning was the Tennessee Valley Authority (T.V.A.), formed in 1933. Even the great success of this project of comprehensive riverine management was not enough, though, to eliminate narrow decisions and lead to proper development of entire watersheds. The dams constructed on other river systems in imitation of the T.V.A. lacked even the fish ladders to assist anadromous fish to spawn that had been spoken of as necessities by Marsh, Baird, and G. B. Goode. Pork-barrel politics, the constant refrain of fiscal austerity, and cost-effectiveness prevented adoption of the proven technology of fish ladders on the major salmon streams of the Pacific Coast, for instance. [40]

[37] Ibid., p. 210; Leopold, *Round River*, p. 159.

[38] Leopold, *Round River*, p. 165.

[39] Harold Gilliam, "The Fallacy of Single-Purpose Planning," *Daedalus* 96 (Fall, 1967): 1142–57.

[40] Ray Palmer Teele, *The Economics of Land Reclamation in the United States*, pp. 134–36, 310–11, 323–28.

After certain natural catastrophes, like floods or dust bowls, environmental policies were more likely to be passed. But even then, it was easier to get constituencies, lobbyists, representatives, and bureaucrats to overcome special interests on limited single-purpose programs—such as soil conservation, flood control, or navigational improvements—than to offer comprehensive programs. The narrowness and short-sightedness of Depression-era programs to reclaim worn-out lands provided a case in point. "Government bought land," Leopold wrote. "All this, once the CCC camps were gone, was good, . . . but not so . . . the maze of new roads that inevitably follow governmental conservation. To build a road is so much simpler than to think of what the country really needs. A roadless marsh is seemingly as worthless to the alphabetical conservationists as an undrained one was to the empire-builders." [41]

Having himself worked for the U.S. Forest Service, Leopold expressed a still more fundamental concern over the efficacy of institutional protection of nature, which was based on his years of experience in government. He believed that most "growth in governmental conservation is proper and logical," some even inevitable. He was concerned, though, that the "clear tendency in American conservation to relegate to government all the necessary jobs that private landholders fail to perform" would result in a governmental responsibility that simply could not be met, financially or otherwise. Government could not indefinitely afford to rectify the mistakes of the private sector, as more and more technological interference with healthy landscapes left eroded and marginal lands for the commonweal. Nor could government always see clearly enough to formulate the necessary policies. [42]

His solution to this inescapable dilemma was, again, individual ethics. He relied on a wilderness aesthetic, drawn from the American naturalist tradition, to foster individual responsibility for protecting biotic integrity and reviving the land's health. His ecological credo required protection of all functioning parts of the landscape as essential to successful and benign environmental use by humans. Like Thoreau's, Leopold's appreciation for the interdependence of species within the biotic community led him to champion the preservation of marshes. His advocacy of wilderness preservation through applied ecology helped to establish the essential role of marshlands in the economy of nature.

[41] Leopold, *Sand County Almanac*, p. 107.
[42] Ibid., pp. 249–50.

Furthermore, Leopold warned that even if some "tinkering" were theoretically allowable, science comprehended physical laws too inadequately to know where such changes were acceptable. His ideas reflected a major intellectual shift occurring within the scientific community, as relativity and probability replaced absoluteness and certainty in the formulation of laws and hypotheses and the reporting of experimental results. Leopold joined others in suggesting that the scientist "knows that the biotic mechanism is so complex that its workings may never be fully understood." [43] How, then, could scientists give ecological advice to government? Much of the internal debate about the limitations of science was lost on the public, whose faith in technology had been reinforced by the successful war effort, but Leopold insisted that "the outstanding scientific discovery of the twentieth century is . . . the complexity of the land organism. Only those who know the most about it can appreciate how little is known about it." He concluded: "If the biota, in the course of aeons, has built something we like but do not understand, then who but a fool would discard seemingly useless parts? To keep every cog and wheel is the first precaution of intelligent tinkering." [44]

Despite the urgency of Leopold's ecological imperative, his land ethic faced two serious challenges in the postwar world. First was the cultural disposition, long ingrained in Americans, to treat land as a real estate venture ensuring upward mobility. The second stemmed from the apparent inverse relationship between the law of the ecological tithe and the marginal value of land. Regardless of the particular circumstances, every time energy is transformed from one form to another an overall net loss of usable energy occurs. Physicists call this increasing entropy or randomness in a system. Because of this loss, efficiency declines, and systems convert less usable energy into more readily utilized forms. More and more wildness is required to maintain a fund of energy producers and to replenish the earth's energy potential. Energy availability decreases as wild lands, especially marshlands, are disturbed and "developed."

The marginal value of land, on the other hand, tends to increase in proportion to the demand for housing and cultural affluence, especially as the population grows. While a long-term balance may be necessary between wilderness, or at least farmland, and residential or industrial property, the short-term economics of scarcity drives up the price of land, and

[43] Ibid., p. 241.
[44] Ibid., p. 190.

the demand to convert economic "wastelands" is not necessarily countered by the recognition that it results in the destruction of biological integrity.

The problem is compounded by the dependence of local government on property taxes. As the value of property increases, so does the tax, which encourages owners of farmlands and "wastelands" alike to convert their lands to revenue-producing uses—that is, to engage in land development—in order merely to pay the taxes. Policies that tend to increase demand or add more buyers in the marketplace exacerbate this situation. Availability of V.A. and F.H.A. financing after the war had just such an effect, providing federal incentives for both urban and suburban home buying. The federal highway plan added to the price spiral in two ways. First, the construction of roads into suburban areas brought both the center city and the once-isolated coastal regions within commuting distance for suburban dwellers. Second, the confiscation of acreage for constructing superhighways removed valuable farmland, wasteland, or residential properties from the market and the tax roles. Thus the price (and the proportional tax burden) of remaining private property increased, and the incentives for developers to convert farmland and wasteland into new residential property rose accordingly.[45]

Eventually the question arose: Who could afford Leopold's ethics? If the taxes on the owner of marsh acreage encouraged the subdivision and sale of the property, only a wealthy owner could follow the land ethic's injunction. Some taxes are assessed based on the property's highest potential use, and under such circumstances an owner would lose economic capital all the while he was storing up eco-energetic capital by maintaining a healthy tidal marsh. Although a rising clamor for estuarine protection has placed tidal marsh preservation on the political agenda of the nation, the inherent conflict between ecology and economics remains unsettled.

Leopold recognized the economic disincentives that hindered the development of his land ethic. He complained that "land-use ethics are still governed wholly by economic self-interest." In the attempt to counter such narrow views of the human obligation, he called on biblical tradition: ". . . Ezekiel and Isaiah have asserted that despoliation of land is not only inexpedient but wrong. Society, however, has not yet affirmed their belief." Rather, he wrote, "man always kills the thing he loves, and so we the

---

[45] U.S. Department of the Interior, Fish and Wildlife Service, *National Estuary Study*, VI, 1–8 (Appendix 1: "Effects of Tax Laws and Policies").

pioneers have killed our wilderness." Through his writings, the next generation of naturalists and scientists were influenced by Leopold's veneration
of healthy landscapes. In closing his notes on the deltaic estuary of the Colorado River he remarked, "I am glad I shall never be young without wild
country to be young in." After all, he asked rhetorically, "of what avail are
forty freedoms without a blank spot on the map?" [46]

Aldo Leopold exhibited a religious devotion to preserving necessary
natural enclaves. Such islands of diversity amidst a sea of sameness assured future human survival because the biological integrity of wild sanctuaries could sustain ongoing coevolution of plant and animal associations.
Without healthy biotic communities, the human capacity for successful
cultural adaptations to future environmental changes would be curtailed
and the human experience of diversity diminished. Sooner or later every
culture must instill the past generations' collective wisdom, acquired from
countless nature studies, in the education of the new generation if order,
civility, humaneness, and a dedication to landscape renewal were to persist. When he died, fighting a neighbor's brush fire on the anniversary of
John Muir's birth in 1948, Leopold left to the next generation a staggering
agenda for the social change of "attitudes and implements." [47]

[46] Leopold, *Sand County Almanac*, pp. 157, 158, 239, 245.
[47] R. Nash, "Aldo Leopold," *Dictionary of American Biography* (New York: Scribner's
Sons, 1974), p. 484.

7

Estuaries and the New Ecology

When we go down to the low-tide line, we enter a world that is as
old as the earth itself—the primeval meeting place of . . . earth and
water, a place of compromise and conflict and eternal change. For us
as living creatures it has special meaning as an area in or near which
some entity that could be distinguished as Life first drifted into
shallow waters—reproducing, evolving. . . . We come to perceive
life as a force . . . strong and purposeful, as incapable of being
crushed or diverted from its end as the rising tide.

—Rachel Carson, 1955

THE expansion of wisdom is a difficult task for an individual, let alone a
society. In democratic cultures wisdom often may be overlooked at the voting booth and in the formulation of policy. For citizens and civil servants
alike, confronting popular notions of efficacy and justice takes courage.
Rachel Louise Carson was an advocate of unpopular causes, including fish
and bird protection, preservation of wild seashores, and opposition to
heedless use of "miracle" technologies. Her final, best-known attack on
the indiscriminate uses of DDT as an insecticide was only one part of her
larger and more positive view of the world and humanity's place in the eco-
energetic scheme. In her writings she concentrated on the poetic truths reflected in the research of marine biologists. More than any other scientist
she popularized the oceans and the shores, with a profoundly religious
respect for nature and a practical appreciation for human ignorance.[1]
A contemporary of her fellow preservation ecologist Aldo Leopold, Carson focused more explicitly than he on oceans and shores.

She wrote in the tradition of Adam Seybert, William Cullen Bryant,
Henry David Thoreau, and John Muir concerning the necessity of marshes,

---

[1] Paul Brooks, *The House of Life: Rachel Carson at Work*, pp. 1–50; Donald Fleming,
"The Roots of the New Conservation Movement," *Perspectives in American History* 6
(1972): 11–14, 23–24.

the beauty of the seashore, and the intrinsic value of wild things. When she began writing in 1937 about the sea and its creatures, these sentiments were shared by few powerful people. Concern for wildlife had taken a deservedly low priority during the worst economic depression in recent world history. Cape Hatteras had just been preserved as the nation's first seashore by the federal government under the marginal lands program, but the human household was in disarray, and reclamation, not preservation, motivated what federal tidelands policy there was. As part of solving domestic problems, some federal planners argued that worn-out lands should be abandoned and that the farmers who owned them required a national resettlement program. That program included North Carolina's Outer Banks, where Edmund Ruffin had early recognized the need for comprehensive reclamation policies. But until 1961, with the creation of Cape Cod National Seashore, the rehabilitation of the coast by federal agencies languished.[2]

Exploring the relationship of Rachel Carson's works to the national mood after the Second World War gives a clearer picture of popular support for coastal conservation. She advocated the preservation of the ecological integrity of seashores on scientific, aesthetic, and practical grounds, but the popular desire for recreation demanded access to critically sensitive marine environments. While these impulses eventually came into conflict, both strains were evident by the early sixties in the clamor for a new kind of federal program to safeguard the seashore and its vulnerable estuaries. Rachel Carson was not solely responsible for the emergence of coastal wetlands preservation as a political issue, but she did give the finest expression in American letters to the scope and meaning of an estuarine preservation ideal. Her popularizations of the sea reflected the research of Gordon Gunter, Joel Hedgpeth, Paul Galtsoff, and Eugene Odum, among many others. People listened to and read what Rachel Carson had to say and were led to defend what our ancestors had always taken for granted: the resiliency of our inland seas.[3]

Born on May 27, 1907, in Springdale, Pennsylvania, Rachel Carson grew up on a small farm, then graduated from the Pennsylvania College for

[2] Rachel Carson, "The Undersea," *Atlantic Monthly*, 1937, reprinted in Brooks, *House of Life*, pp. 25–35; "Rim of the Sea Becoming Crowded and Dirty," in *Man's Control of the Environment*, ed. Jack Nease, pp. 21–31.

[3] Phillip Sterling, *Sea and Earth: The Life of Rachel Carson*, pp. 112–13, 119–23, 139, 150–61; Fleming, "Roots of the New Conservation," pp. 13–14, 28–29.

Women in 1928. She received an M.A. from the Johns Hopkins University in 1932 after summer studies at the Woods Hole Oceanographic Institution in Massachusetts. She joined the Bureau of Fisheries, then part of the Department of Commerce, in 1936. That department, under the directorship of Herbert Hoover from 1921 until his election as president in 1929, had taken the lead in resource policy making and the protection of the Alaskan salmon industry. In addition Hoover had taken an interest in establishing fish nurseries and controlling water pollution and flooding through state and federal cooperation. While the approaches of the Hoover and Roosevelt administrations were far from ecologically oriented, they did establish the tradition of planning and intergovernmental cooperation. In this milieu Rachel Carson began writing about Chesapeake Bay, coastal wildlife, and seashore preservation.[4]

In an article appearing in 1936 in the *Baltimore Sun* magazine supplement concerning the migrations of shad, Carson suggested that recent declines in the fishery were traceable to "destructive methods of fishing, the pollution of waters by industrial and civic wastes and the development of streams for water power and navigation." Each of these interferences with the life-sustaining energy along the coast had, historically, further depleted the protein-rich fisheries. Marsh and Baird had reiterated Mitchell's earlier warnings of fish declines due to improper fishing techniques. Forbes had joined these critics in chastising polluting manufacturers, while Thoreau had described the natural diseconomy of Billerica Dam on the Merrimac River. Taken separately, each disclimax weakened estuarine resiliency, but together their combined influences exceeded the simple sum of each disturbance and were powerful enough to disrupt the fish nurseries of the Chesapeake Bay. Carson, like Leopold, understood that the federal government was guided by the same prejudices and fiscal strictures as the laissez-faire industrialist had been. With the approval of the Boulder Canyon Project in 1928 and the creation of the Tennessee Valley Authority five years later, federal advocacy of riverine hydroelectric development brought expanded responsibilities for fisheries, water quality, recreation, and public health.[5]

With this widening of the federal role in interstate commerce and nav-

[4] Sterling, *Sea and Earth*, pp. 12–13, 66–83, 86–87; Donald Swain, *Federal Conservation Policy: 1921–1933*, pp. 160–70.
[5] Swain, *Federal Conservation Policy*, pp. 169–70.

igation, serious difficulties were bound to arise. In response to the pollution of the Potomac River, Congress passed the Fish and Wildlife Coordination Act in 1934. The sentiments expressed in the legislation were honorable. Besides seeking interagency cooperation, the act encouraged consulting the Bureau of Fisheries and the Biological Survey when impoundment of rivers was planned by either the Army Corps of Engineers or the Bureau of Reclamation. However, the act had no enforcement provisions, and no money was ever authorized for creation of the fish hatcheries, migratory bird refuges, and fish ladders suggested by the bill to mitigate the effects of water impoundment. As the depression deepened, soil conservation and flood control emerged as overriding federal resource priorities.[6]

Similarly, during the 1920s, city planning had evolved into regional planning because of the chronic need for fresh water, sewage disposal, and parks for recreation. Two new technologies broadened the effects of urbanization: long-distance energy transmission and the automobile spread industry and residential suburbs into rural areas and recreation seekers into remote places. The federal highway program built a coastal road from Monterey to San Simeon along California's Big Sur coast. Maine and North Carolina seashores became accessible to residents of the growing northeastern urban corridor. Planners raced with technology and population to keep abreast of the urban dweller's basic needs for housing, energy, and transportation.[7]

Despite bipartisan efforts to coordinate federal, state, and local programs, resistance to centralization remained strong in the Army Corps of Engineers, and the Department of Agriculture refused to support a push by Secretary of the Interior Harold Ickes for a department of conservation. Ickes further disrupted the surface harmony between federal and state officials when he claimed an overriding federal concern for resources in the submerged lands of the Continental Shelf. While this was a natural outgrowth of federal oil-conservation policies under William Howard Taft and Herbert Hoover, the coastal states were not inclined to acquiesce. Once the war ended, the conflict between federal and state responsibilities in the coastal zone was renewed. The effects of these intergovernmental and in-

---

[6] Arnold Reitze, ed., *Environmental Planning: Law of Land and Resources*, pp. 2–57.
[7] Mel Scott, *American City Planning*, pp. 183–99.

terbureau rivalries were not lost on Rachel Carson, who went with the Bureau of Fisheries to the Interior Department in 1939 and beginning in 1940 worked for the newly created Fish and Wildlife Service.[8]

Carson was intimately familiar with the secondary role wildlife advocacy played in relation to the many other functions of the Bureau of Fisheries. During her tenure, former secretary Harold Ickes had written a scathing indictment of the more powerful Army Corps of Engineers, which wielded vast political influence, often to the detriment of comprehensive conservation. Just prior to the war in 1941, Carson's first book on marine ecology appeared, entitled *Under the Sea Wind* and patterned after the British best-seller *Tarka, the Otter* (1927) by Henry Williamson. Based on an article that had appeared in *Atlantic Monthly* in 1937, it was written from the perspective of the animals inhabiting or frequenting estuaries and beaches. As the rivalry between state and federal governments for the tidelands sharpened, Carson's book gained added significance.[9]

Following the war, Carson was made editor-in-chief of the Information Division of the Fish and Wildlife Service, with a staff of six. Her pamphlets on fishes and a series called "Conservation in Action" were highly successful. Since 1918 the federal government under treaty with Canada and Mexico had been obligated to create migratory bird refuges, and Carson's series was designed to inform the public of the need for such preserves. In the introduction to the series, she noted that "wild creatures, like men, must have a place to live." However, she warned, "as civilization creates cities, builds highways and drains marshes it takes away . . . land . . . suitable for wildlife." She then made for the layman the crucial connection between habitat and population expansion: "As their spaces for living dwindle, the wildlife populations themselves decline." The pamphlets, of which she wrote four, appeared between 1947 and 1950. Her introductory remarks dovetailed nicely with the sentiments of Aldo Leopold, first brought to the attention of the public in 1949, with the posthumous publication of *A Sand County Almanac*. Both Carson and Leopold had broadly stated the underlying tension between human living space and wilderness

[8]Harold Ickes, *The Secret Diary of Harold LeClare Ickes* (New York: Simon and Schuster, 1953), II, 127, 311, 330, 493–96; Arthur Maass, *Muddy Waters: The Army Corps of Engineers and the Nation's Rivers*, pp. i–ix, 3–24.

[9]Fleming, "Roots of the New Conservation," pp. 13–14; Sterling, *Sea and Earth*, pp. 87–88, 91–95; Brooks, *House of Life*, pp. 22–30.

preservation that would characterize the next twenty-five years of conservation debates.[10]

The public acceptance of Rachel Carson's books in the 1950s was only part of a larger national interest in science and technology, initiated by concerns about military security. A new synthesis of ideas concerning humanity's role in nature was gradually coalescing. The postwar national temper assured that Carson's books were not isolated in the intellectual development of an ecological tradition. Environmentalists like Fairfield Osborn and Harrison Brown suggested that the central issue for the remainder of the century was the finite amount of world resources in relation to the exponential growth of the world's population. Thus the demand for food, fiber, and energy would continue to drive up the price of basic life-sustaining commodities, as the quality of the biotic communities—their ability to store eco-energetic capital—declined. Another group of ecologists grew out of the revolt of biologists and physicists against the uses of nuclear energy. Barry Commoner led the political assault of those who believed that nuclear technology was the central problem of the age. The nuclear debates concerning the hydrogen bomb, atmospheric testing, and atomic fallout kept scientific questions before the public and convinced publishers of the demand for readily understandable books on highly complex scientific research. The implications of nuclear development for the oceans began to be apparent in the late fifties, when oceanographers warned of the danger of storing spent nuclear fuels or radioactive wastes beneath the sea.[11]

Carson, in the meantime, was slowly gathering data for a book on oceanography. She wanted this second work to be "a book for anyone . . . that will be easily understood and imaginatively appealing to the reader untrained in science." As with her later works, before the Oxford University Press edition was released in July of 1951, parts of *The Sea Around Us* appeared in the *Yale Review* and *New Yorker Magazine*. By November,

[10] Sterling, *Sea and Earth*, pp. 99–100.

[11] Fleming, "Roots of the New Conservation," pp. 42–49, 52–63; Fairfield Osborn, *Our Plundered Planet* (Boston: Little, Brown & Co., 1948); Harrison Brown, "The Dangers We Face," in *The Atomic Age: Scientists in National and World Affairs*, ed. Morton Grodzins and Eugene Rabinowitch, p. 170; William Vogt, *Road to Survival* (New York: Sloane, 1948); Paul Ehrlich, *The Population Bomb* (New York: Ballantine, 1968); Roger Revelle and Milner B. Schaefer, "The Ocean as a Receptacle for Artificially Radioactive Materials," in *Man and the Sea*, ed. Bernard L. Gordon (Garden City, N.Y.: Doubleday, 1970), pp. 236–48. See also pp. 130–34, 296–310, 539–41 in Grodzins and Rabinowitch, *Atomic Age*.

1951, the book had sold more than 100,000 copies, and it remained on the *New York Times* best-seller list for eighty-one weeks.[12]

In accepting the National Book Award for 1951, Carson explained the overwhelming success of her book: "We live in a scientific age. Science is part of the reality of living; it is the what, the how and the why of everything in our experience." The book's popularity placed Carson in a central role in interpreting the new scientific outlook that formed the basis of environmental concern for the next twenty years.[13]

The success of *The Sea Around Us* led to several important events in Carson's career. Oxford Press reissued *Under the Sea Wind* in 1952, and this time it too made the best-seller lists. The two books now freed Carson to write professionally full time. In 1953 she resigned from the Department of the Interior to begin work for Houghton Mifflin on a book on the seashore. She also purchased a summer home in Boothbay Harbor, Maine, and began numerous trips from there to Florida to gather data for the forthcoming *Edge of the Sea*. This third book focused on the Atlantic shore in order to describe the delicate yet resilient life forms that were now of wide interest to her fellow citizens. Red tides, water pollution, and nuclear-waste dumps in the seas had all galvanized experts and laymen alike to explore and describe the relationships among earth, air, and water. Carson had described how the sea as life's first environment was the single most important factor in maintaining life on earth. Reverence for that life and the need for human decency toward wild creatures had been underlying themes in Carson's writings since her early association with the Fish and Wildlife Service. In 1955, with the publication of *The Edge of the Sea*, Carson brought that same keen respect for nonhuman life to the nation's coast.[14]

In her first publication in 1937, Carson had written, "There are comparatively few living things whose shifting pattern of life embraces both land and sea." Such creatures, she informed her readers, dwelt both in "the tidepools among the rocks" and on "the mud flats sloping away from dune and beach grass to the water's edge." Across this tidelands habitat, she observed, "land and sea wage a never-ending conflict for possession."[15]

[12] Sterling, *Sea and Earth*, pp. 105, 109, 111, 113, 114.

[13] Fleming, "Roots of the New Conservation," p. 12; Sterling, *Sea and Earth*, p. 116.

[14] Rachel Carson, *The Edge of the Sea*, pp. vii–viii; Sterling, *Sea and Earth*, pp. 121, 123–32.

[15] "Undersea," *Atlantic Monthly* (1937), cited in Brooks, *House of Life*, p. 23.

Carson saw, experienced, and wrote about the consequences of the American love of the seashore—an affair that was loving the coast to death. During the tumults over the discovery of DNA as the secret of life and the development of fission as the secret of matter and energy, Rachel Carson reserved her excitement and wonder for the life of the sea and the subtler energy conversions facilitated by its wide expanse. She had called the coast "the place of our dim ancestral beginnings." Concerning its ecology, she believed that understanding "comes only when . . . we can sense with the eye and ear of the mind the surge of life beating always at its shores." [16]

In the preface to *The Edge of the Sea*, Carson explained, "I have tried to interpret the shore in terms of that essential unity that binds life to the earth." These diverse coastal wetlands, fought over by local, state, and federal developmental interests, Carson depicted as "crowded with plants and animals." Winds, tides, and fluctuating water quality created a stressful environment along the littoral, and, she recognized, that "in this difficult world of the shore, life displays its enormous toughness and vitality by occupying every conceivable niche." [17]

The herons and deer feeding in the sheltered habitats of the marshes represented the upper layers of the energy pyramid. The foundation niches were occupied by algae, diatoms, or dinoflagellates floating in the waters and transforming sunlight into life. Tiny "insects of the sea" or copepods fed upon these tiny single-celled plants much as grasshoppers preyed on the *Spartina* or cord grass of the lower marsh. The seas also harbored unique creatures for which the land had no counterparts, the filter feeders. Among these varied organisms, none was more commercially significant

---

[16]Carson, *Edge of the Sea*, p. vii.

[17]Carson, *Edge of the Sea*, pp. viii, 11, 36–38. No concept other than succession has been so thoroughly challenged in ecology as that of the niche. Derived from the French word for nest, it holds implications that are easy for naturalists to grasp. But with the coming of energetic analysis to ecology, niche came to mean less an address or occupation of an organism than a four-dimensional space-time potential for transforming energy from one form to another. Part of this process, which allows all life to run counter to the system-wide increase in entropy, is the storage of necessary information. This information concerns the character of environmental changes and the most successful way for an organism to deal with those shifts. Since the turn of the century, geneticists had uncovered the mechanism by which environmental information is stored and passed on to the succeeding generation. Biochemists who discovered DNA and RNA revealed the structural manner of an organism's information storage allowing adjustments to the available niches (Eugene P. Odum, *Fundamentals of Ecology*, pp. 234–39; James D. Watson, *The Double Helix* [New York: Atheneum, 1968], pp. 30–70).

than the oyster. "Adult oysters live in bays, or sounds or river estuaries," wrote Carson, "rather than in water of full oceanic salinity." She explained how the oyster's life cycle had to adjust to the physical demands of the tides. "The larval oyster . . . uses the flow of the tides to carry it into areas favorable for its attachment." Oysters, like mussels and clams, feed by straining out detritus, plants, or animals from the water filtering through their shells. These filter feeders remove and concentrate the vital nutrients trapped by the brackish merging of fresh and salt water. Each of these differing organisms occupies an exclusive niche in the estuarine ecosystem. The niche represents the favorable interstices within which environment and evolution together permit organic exploitation of the ecosystem.[18]

Concerning the shore zone especially, Carson wrote, "Each time I enter it, I gain some new awareness of its beauty, and its deeper meanings, sensing that intricate fabric of life by which one creature is linked to another and each with its surroundings." Like Aldo Leopold, who grasped the enormity of life without quantum mechanics, Rachel Carson without specialized investigations of atoms and chemistry understood "the whole life of a creature" and "its relations to the particular sea world in which it lived." Overlooking the disputes within the scientific community, Carson cut to the simple heart of the matter. In her writings she stressed that "the shore has a dual nature . . . belonging now to the land, now to the sea." She depicted this interplay within the ecosystem of an estuary, where the terrestrial and oceanic worlds were brought into intimate relation by the daily actions of the tides.[19]

The success of Carson's three books on marine ecology reflected the enormous boom in postwar oceanographic research. Wartime research had been primarily concerned with beaches and navies. Many oceanographers continued their close ties to the government once the conflict ended. Carson was merely the most visible of the marine world's many investigators. Terrestrial ecologists also helped to alter significantly the scientific understanding of the seashore's least appreciated parcels, the marshes. Eugene Odum, the director of the University of Georgia's Ecology Institute, later recalled, "It was in 1954 that I first really saw the Georgia barrier-island

    [18]Carson, *Edge of the Sea*, pp. viii, 11–12; Rachel Carson, "Our Ever Changing Shore," *Holiday Magazine*, July, 1958.
    [19]Rachel Carson, *The Sea Around Us*, pp. vii–xiv. The work itself was dedicated to Henry B. Bigelow, director of the Woods Hole Oceanographic Institution and marine biologist at Harvard, who was Carson's friend.

salt-marsh estuary." In that year, before the publication of *The Edge of the Sea*, Odum "in the company of colleagues and students . . . peeked into the insides of the estuarine system for the first time." The occasion was an invitation from R. J. Reynolds, Jr., for the university to establish a research laboratory on Sapelo Island. As they motored along the creeks, Odum explained, "The notion came to us . . . that we were in the arteries of a remarkable energy absorbing natural system whose heart was the pumping action of the tides." Odum came to realize that "the entire tideland complex of barrier islands, marshes, creeks and river mouths was a single operational unit linked together by the tide." [20]

Eight years before Odum's close work with estuaries began, an important study by Claude Zobell of the Scripps Institution of Oceanography had appeared that would crucially support Odum's findings. In *Marine Microbiology*, Zobell had distilled his own and other pioneering efforts on the role of organic detritus in the ecology of the seas. As Odum remarked, this "particular organic matter rich in vitamins and calories," detritus, is usually found in what "most people would call mud." Estuaries had remarkably high concentrations of organic detritus sustaining high levels of productivity in organisms that utilized it for food, such as clams, oysters, barnacles, or mussels. Even as Odum's team of specialists and graduate students attempted to gauge the productivity of tidal mashes, other marine ecological research was contributing to a new understanding of estuaries and their significance in fisheries. [21]

In 1957 the first compilation of important work on sea life from an ecosystems perspective was published, under the general editorship of Joel W. Hedgpeth and entitled *Treatise on Marine Ecology and Paleoecology*. It contained more than twelve hundred pages of studies from worldwide oceanographic experts. Hedgpeth himself authored sections on the history of marine ecology and estuarine classifications, among other articles. In 1952 Hedgpeth had revised a leading popular guide to Pacific inter-

[20] Robert Hanie, *Guale, the Golden Coast of Georgia*, pp. 19–21.

[21] Ritchie Ward, *Into the Ocean World: The Biology of the Sea*, pp. 230–40; Odum, quoted in Hanie, *Guale*, pp. 19–24; Eugene Odum, "Description and Productivity of Georgia Salt Marsh Estuaries," manuscript, p. 7; Eugene Odum, "The Pricing System," *Georgia Conservancy Magazine* (4th quarter, 1973): 8–10; Odum, *Fundamentals of Ecology*, pp. 352–62. The first conference on estuarine research was hosted by the Sapelo Island Marine Station on the Georgia Sea Islands in 1964, and its findings were published by the American Association for the Advancement of Science (George Lauff, ed., *Estuaries*, A.A.A.S. Publication No. 83, 1967).

tidal organisms, *Between Pacific Tides*, originally published in 1939 by Edward Ricketts and Jack Calvin. Hedgpeth had studied at Berkeley under Solomon Felty Light during the war. Concerned over the lack of fishways on major hydroelectric dams in the west, young Hedgpeth had formed a one-man society for the prevention of "progress" in the mid-forties.[22]

The work of Joel Hedgpeth concerning the nursery function of estuaries for anadromous fish like salmon, trout, and transplanted sea bass brought the important economic function of estuaries to the closer attention of the public. The marshes along the Pacific flyway originally had supported millions of migratory waterfowl, and the creeks beside the mud flats where the birds fed had supported commercial fish. Flounder, crabs, and shrimp, in addition to oysters and anadromous fish, required the fertile waters of estuaries. The salinity variation within estuaries was crucial to the ecosystem of San Francisco Bay, and Hedgpeth had frequently been asked how to protect the wooden piers of the harbor from attacks by ship worms.

These "worms," actually invertebrates related to boring clams, drill into lumber pilings and thrive under conditions of oceanic salinity. Coastal settlement, especially in the arid regions, taxed the freshwater capacities of coastal aquifers. Upstream diversion of water for reservoirs and farming had the ironic effect of providing the great valley with more water while depriving the San Francisco Bay–San Joaquin Delta system of flushing by fresh water. Without the seasonal flushing and the year-long storage of fresh water in the marshes of the Sacramento–San Joaquin Delta, the salinity of the bay increased. In short, this form of disclimax was triggered by salt-water intrusion. Especially in the dry season, the removal of fresh water from the bay, the marshes, and the underground aquifers encouraged the landward seepage of the more saline ocean water. The habitat of the marine boring clams therefore expanded, their numbers increased, and piers previously free from their infestation weakened and collapsed.[23]

Since one-fourth of all the liquid fresh water in the biosphere exists in North America, the problems of water quality and conservation were long

[22] Joel Hedgpeth, "Voyage of the Challenger," *Scientific Monthly* 63 (September, 1946): 144, 194–202; Joel Hedgpeth, *Treatise on Marine Ecology and Paleoecology*; Joel Hedgpeth, *Between Pacific Tides*, rev. ed., pp. v–x.

[23] Joel Hedgpeth, *Seashore Life of the San Francisco Bay Region* (Berkeley: University of California Press, 1962); Hedgpeth, *Between Pacific Tides*, pp. 231–77, 307–46, 358–60, 370–73, 515, 518–21.

ignored. But as the population expanded, the per capita use of water vastly increased, and even the tropical coral shores of south Florida were not immune from saltwater intrusion. Along the entire extent of the coast, where wells tapped groundwater supplies, the problem was the same. Sedimentation and reclamation threatened estuarine ecosystems, and the intrusion of salt water was an added problem.[24]

The most insidious threat to the nation's estuaries, though, appeared in the form of water pollution. Organic wastes, which can be well degraded by marsh ecosystems, were targets of Progressive reform at the turn of the century. Water purification to remove bacteria from urban water supplies was a technology used widely after the war. But several other types of pollution, including viral infection and biological nutrient enrichment from nitrate fertilizers or phosphate detergents, still threatened river and coastal waters. Synthetic chemicals including pesticides, plastics, and radioactive wastes were new contributors to water pollution. Finally, thermal pollution, or the artificial heating of water by electrical generating plants, and oil spills reflecting the increased per capita demands for energy polluted aquatic and marine resources.[25]

The effects and extent of tidal-marsh destruction on the energetic potential and productive capacity of estuaries was unknown until 1961. Research conducted then along the lines of Lindeman's study of a cedar bog lake revealed the incredible efficiency of salt-marsh vegetation in converting captured sunlight into matter. The plants in Lindeman's study garnered .1 percent of the solar radiation, compared with 6 percent of the available solar radiation obtained yearly by tidal-marsh plants. The enormous biotic potential of an acre of Georgia's tidal marshes was more than ten tons of dry-weight organic matter per year. This productivity was five times as great as that of an acre of corn. Intensive European farms produced only seven tons of crop per year. Odum's studies demonstrated that some wetlands of the Atlantic Coast were some of the richest biological communities on the face of the earth. No longer could their potential be viewed

[24] R. Reinow and L. Reinow, *Moment in the Sun* (New York: Ballantine/Sierra Club, 1967), p. 69.

[25] Fleming, "Roots of the New Conservation," pp. 7–15, 40–51, 82–91; Anne W. Simon, *The Thin Edge: Coast and Man in Crisis*, pp. 81–120; Michael J. Barbour, Robert B. Craig, Frank R. Drysdale, and Michael T. Ghiselin, *Coastal Ecology of Bodega Head* (Berkeley: University of California Press, 1974), pp. 211–48; Institute of Ecology, *Man in the Living Environment: A Report on Global Ecological Problems*, pp. 220–67; Wesley Marx, *The Frail Ocean*, pp. 59–110, 155–79.

simply in pejorative terms. Beneath the muck and mire of the tidal marshes, scientists had uncovered the closest candidate for the status of a natural garden. Fed the land's nutrients by rivers, plowed by the tides of the sea, and mulched by the decay of *Spartina* grasses, this estuarine garden provided the necessary fuel to sustain both marine and terrestrial life.[26]

Odum's researchers at the Sapelo Island Marine Station, amid the Georgia sea islands, offered proof of the estuarine ecosystem's primary productivity. In the summer of 1961, Odum's published report of their findings stated, "We see that estuaries tend to be more fertile than either uplands on the one hand or the sea on the other." He identified three reasons for the fertility of the ecosystem. The first was the salinity gradient from marine to fresh water that, together with various organisms, acted as a trap to hold within the estuary the necessary nutrients for plant growth. Second was the action of the tides in creating a flowing system bringing in the necessary oxygen and removing the harmful wastes, much as the circulatory system did in the body. Finally, their cooperative research efforts had demonstrated that three distinct productive sectors were brought into play at different seasons, thereby maintaining the year-round productivity of the habitat. The *Spartina* grass, growing as rapidly as sugar cane about the edges of the creeks and mud flats—and secreting its own insect repellent—was the first unit of production. Two crops of cord grass or *Spartina* per year were the average yield in these particular southeastern estuaries. Growing profusely on the banks and mud flats and suspended in the water, the second productive unit, which formed "a beautifully adapted community" year-round, was mud algae. The last significant unit also floated in the water but bloomed only twice a year—the phytoplankton, shared by the estuary and the sea as a basic component of primary productivity.[27]

Odum recognized the reason this ecosystem's richness had not been readily apparent to either the scientist or the laity: "More human food per acre . . . is obtained from a wheat field than from the more fertile estuary because only a small fraction of the latter production reaches the human link in the food chain." This criticism, part of the outlook of the "new conservation," considered the pecuniary value of a bio-geochemical cycle. Simply stated, it asked, what is the value to human society of maintaining a healthy ecosystem supporting a significant number of organisms below

[26] Eugene P. Odum, "The Role of Tidal Marshes in Estuarine Production," *Conservationist* (June–July, 1961): 12–35.

[27] Ibid., pp. 13–15.

us on the food chain by trapping sunlight and the necessary biological nutrients?[28]

"Unfortunately too many so-called conservationists and engineers also view these estuarine environments as would a dry land farmer," Odum lamented, adding his voice to the debate over the social value of unrestrained reclamation. He concluded that "we have plenty of land for crops without converting potentially more useful wetlands." Mariculture, or marine farming, was a more mutually beneficial use for this fragile biome. Odum insisted that the estuary "must be considered as one ecosystem or productive unit." Therefore, in the management of this habitat, natural energy subsidies and cycles should be utilized for the overall benefit of the ecosystem rather than the maximization of production as pursued in monocrop agriculture. Finally, Odum borrowed Hugh Hammond Bennett's notion that had had great success in restoring the health of the land during the New Deal. He suggested "something akin to the soil conservation district program which involves the voluntary cooperation of private and governmental interests on a large scale." He added a warning, though, that "the 'agronomic' approach employed in land management must be considerably broadened to avoid serious mistakes." In stressing the need for biological engineering in the human use of the estuary, he echoed the New Deal belief that "some sort of unified planning is overdue." [29]

The genesis of Odum's call for comprehensive planning went back at least to 1929 and his father's work on President Hoover's Research Committee on Social Trends. Howard W. Odum, the father of both Eugene and Thomas Odum, an energy ecologist, was a sociologist at the University of North Carolina and author of the influential book *Southern Regions*. Howard Odum was part of the broad tradition in regional planning that was represented by the Regional Planning Association of America. This was a loose organization of "broad social vision" gathered around Governor Gifford Pinchot of Pennsylvania that included Benton Mackaye, Lewis Mumford, and Henry Wright. Many of their ideas were later reflected in the comprehensive plans of the Tennessee Valley Authority in 1933. Later, Howard Odum served on the North Carolina counterpart of the New Deal's National Planning Board, also established in 1933. But these people represented merely one approach to planning—a rather politically unrealistic

[28] Ibid., pp. 14, 15; Fleming, "Roots of the New Conservation," pp. 7, 64–73.

[29] Odum, "Role of Tidal Marshes," pp. 14, 15, 35; Stewart Udall, *The Quiet Crisis*, pp. 156–58.

approach considering the vicissitudes encountered by comprehensive planning after the First Hundred Days.[30]

The general weaknesses of regional planning mirror those of Eugene
Odum's suggestions for estuarine conservation districts. The initial strength
of regional planning sentiment dissolved in the face of war and urban
growth. Short-term economic considerations acted as deterrents to truly
area-wide regional policy formulation that would have balanced open-
space preservation with the need to develop urban industry, housing, and
recreational needs. The older constraints of clean water, sanitation disposal, and transportation continued to hamper the formulation of policies
that were acceptable to local governments, states, and federal bureaus. At
the very heart of the planning problem was the fact that no constituency
existed to turn the formulations of city or regional planners into workable
political programs. Plans accumulated on the shelves rather than being
used as blueprints for the future construction of any region, let alone the
nation.

From its inception the regional planning movement has had an urban
bias because of demographic and economic realities. Statistically speaking, over half the U.S. population lived in urban settings of more than
twenty-five hundred inhabitants after 1920. Generally, the only employ-
ment for engineers and planners was with municipalities of ten thousand or
more and in large metropolitan areas. Although Benton Mackaye spoke of
the need for rural areas and the country-life movement of the turn of the
century, furthered by Liberty Hyde Bailey, had also addressed the needs of
nonurban America, they were exceptions. As the automobile and long-
distance electrical transmission allowed cities to expand, city planning
turned into regional planning during the twenties. New York drafted a regional plan in 1921, headed by Frederick Delano. The following year Los
Angeles City and County, led by Gordon Whitnall, pioneered regional
planning out west, and the Delaware River Valley plans under the aegis of
Philadelphia were launched in 1924.[31]

[30] Scott, *City Planning*, pp. 191, 274–76, 304, 367. For example, Gifford Pinchot
wrote in 1910: "The first principle of conservation is development. . . . Conservation stands
emphatically for the development and use of water power now, without delay. It stands for the
immediate construction of Navigable Waterways under a *broad and comprehensive plan* as
assistance to the railroads" (Pinchot, *The Fight for Conservation* [1910, reprint ed., Seattle:
University of Washington Press, 1967], pp. 43–44, emphasis added).

[31] Scott, *City Planning*, p. 184; Andrew Denny Rodgers, *Liberty Hyde Bailey: A Story
of American Plant Sciences* (Princeton: Princeton University Press, 1949), pp. 351–80; Benton Mackaye, *The New Exploration: A Philosophy of Regional Planning*, quoted in Roderick
Nash, ed., *The American Environment*, pp. 99–105.

The Philadelphia plan addressed many familiar problems, such as river pollution, improvement of water supplies, coordination of transportation facilities, and preservation of recreational areas. They also included an innovation, the use of technical advisory committees to bring the special expertise of economists, social scientists, architects, engineers, and behavioral scientists to bear on planning questions. The problems encountered by Philadelphia were also indicative of planning's future difficulties. Lack of agreement on the precise definition and therefore breadth of regional planning was an underlying weakness in Philadelphia, as well as other parts of the country.

Henry Wright, as the New York State Commissioner of Housing and Regional Planning, broadly defined the purposes of regional planning in 1926. As Gifford Pinchot had, Wright saw regional planning as including "the conservation and the future development of the resources of a region to the end that an economic gain may not involve inevitable social loss." To achieve this he suggested "the preservation of all existing natural values both of the country and of the city." And despite the controversies in Los Angeles over possession of Owens Valley water rights and in San Francisco over inundation of the Hetch-Hetchy Valley for municipal water and power, Wright reiterated that regional planning "does not mean the complete subordination of country to urban influences." Yet the difficulty concerning coastal wetlands was exactly what Wright had preached against— urban dominance. Indeed, preservation of relatively undisturbed wetlands lost out to both agricultural planning, as Odum suggested, and to urban planning.[32]

A further stumbling block to the use of comprehensive planning as an estuarine-preservation tool was the regional concept's poor definition. Few if any of the engineers and architects who staffed or advised most planning departments would have practically agreed with Lewis Mumford's definition of regionalism, shared by Wright and others. Mumford, in an address delivered on May 1, 1925, described a region as "being any geographic area that possesses a certain unity of climate, soil, vegetation, industry and culture." Within this area "the regionalist attempts to plan," Mumford felt, "so that the population will be distributed . . . to utilize, rather than to nullify or destroy its natural advantages."[33]

Mumford's conception is paralleled by Charles Elton's (1927) notion

[32] Scott, *City Planning*, p. 221.
[33] Ibid., pp. 198–219.

of the biotic community and Leopold's (1933) stress on carrying capacity of the land organism. Politically speaking, however, both regional and urban planners had to work within a federal system that operated on congressional-district or state-constituency lines that often crisscrossed the kind of regionalism that planners of broad-based land and water use had envisioned. Political power, especially in the malapportioned legislatures of the 1920s and 1930s, was splintered among villages and towns. They opposed any urban influence over their traditional spheres of interest. Locals also fought state and federal programs that sought control over patronage and public-works funding.

The contributions of Mumford, Wright, and others were significant in spite of the tensions among the members of the Regional Planning Association of America, the growing number of architects and engineers who endeavored to create viable plans, and the politicians who sought to further their constituents' interests. Mumford's definition of a region presages Tansley's (1935) notion of ecosystem by encompassing soil, climate, and vegetation patterns and even includes the social system, with emphasis on common cultural and industrial legacies. The association stressed the linking of urban and rural planning for the truly satisfactory formulation of regional plans. Its members also desired to control the use of natural resources and land-use speculation. The most effective tool in this regard was state police powers, but unless the region was within the jurisdiction of a single state, the efficacy of such controls was seriously hampered. Finally, the association emphasized the need for meeting social needs through adequate recreation, transportation, energy, and housing.[34]

Philadelphia and New York were two important estuarine regions where the tri-state character of the regions hindered implementation of broad-based planning. The Philadelphia plan was sponsored by the Russell Sage Foundation and was modelled on the *Regional Plan of New York and Its Environs*. Since the tidal influence on the Delaware River extends above Philadelphia to Trenton, that New Jersey region lies within the river's estuary. The Philadelphia efforts were spearheaded by Russell Van Nest Black, who had been associated with San Francisco planning in 1924–25. His group collected detailed information and accurately predicted that the suburbs around Philadelphia would grow faster than the city proper because of

[34]Donald Worster, *Nature's Economy: The Roots of Ecology*, pp. 294–309; Scott, *City Planning*, 215–20, 223–24, 226, 430, 532–33.

highways and long-distance electrical transmission. He suggested the sequestering of streamsides, hilltops, and wooded hillsides as public lands for the maintenance of watershed. While his plan met stiff resistance from business interests, the educational function of Philadelphia's Tri-state Regional Planning Federation was important. After Black quit, the Great Depression caused the plan to languish in the early thirties. However, the data collected were an important asset for revived planning begun in the same region in 1954 in the Penn-Jersey Transportation Study (later dubbed Penjerdel).[35]

During the New Deal, planning was encouraged by the National Industrial Recovery Act's Title II, under which Harold Ickes created the National Planning Board to advise on Public Works Administration (PWA) projects. The board's work encouraged the creation of state planning boards across the nation, including in coastal states. Russell Black served on the state boards in New York, Pennsylvania, and Virginia, for example. However, the difficulties encountered by Ickes and those favoring national, comprehensive land-use plans were manifold. Older agencies of the federal government were not eager to divest themselves of traditional authority to undertake reclamation, dredging, or public-works construction, let alone to submit those plans to any but congressional scrutiny. Planners would have to cultivate the Departments of the Army and Agriculture if their comprehensive plans were to be translated into effective programs for public works.[36]

The National Planning Board did encourage the drafting of multistate regional plans for the Pacific Northwest and New England designed around the multiple uses of their respective river systems in 1934. The states conferred on problems of water pollution, floods, and erosion, only to incur the ire of the Army Corps of Engineers, which saw its long-exercised powers over these areas threatened. The thirties also saw the publication of a text by Edward Bassett, a leading zoning advocate, entitled *The Master Plan*. In addition to zoning, among the seven elements essential to modern planning Bassett emphasized the need for public reserves of land (this pro-

[35] Scott, *City Planning*, pp. 215–20, 223–24, 226, 304–307, 530, 532–36; Otis L. Graham, Jr., *Toward a Planned Society: From Franklin D. Roosevelt to Richard Nixon*, pp. 24–27, 31–78; Udall, *Quiet Crisis*, p. 157.

[36] Ickes, *Secret Diary*, I, 136, 171–72, 216, 219, 281, 341, 354, and II, 28, 114–15, 132–34. The name of the National Planning Board was changed to National Resources Committee (II, 623, 659, 660, 667–68).

posal was the forerunner of the idea of open-space zoning) and the establishment of pierhead and bulkhead lines to create limits for shoreline filling
and dredging. Still the bias of which Eugene Odum wrote thirty years later
was present: to the engineers coastal wetlands were wastelands or of class-
four land-use potential, encouraging their transformation.[37]

It would take a ground swell of public opinion to change this climate,
and it would be necessary to provide hard statistical data to convince experts that estuarine waters and coastal wetlands were worth preserving for
their biotic potential. Then it would be necessary to salvage protective legislation from unfavorable court reviews. In California, zoning emerged as a
legal precedent for coastal conservation. In 1938, Judge Maurice T. Dool-
ing upheld a Monterey County zoning ordinance protecting the roadside
beauty of the Carmel–San Simeon Coast Highway from laissez-faire development. The court held that the preservation of scenic beauty along un-
usually spectacular highways was necessary to safeguard the aesthetic and
economic assets inherent in "roadside beauty." Yet the disparity between
the rates of change in biological ideas and judicial understanding remained—Aldo Leopold's perception of scenic beauty arising out of the biological integrity and health of the natural community notwithstanding.
The linkage between law and ecology remained tenuous and unpopular at
best. Finally, Rachel Carson placed the preservation of the community's
biological health squarely in relation to the continued survival of human
society. By doing this, her *Silent Spring* facilitated the convergence between biological information and legal interpretation. Her last book reflected the spirit of the times and popularly explained how human institutions ultimately depend on the ecosystem's health and integrity.[38]

The names for the intellectual shift, social agitation, and political legislation facilitated by Carson and many other ecologists between 1959 and
the Arab oil embargo of 1973 have been many. Conservation historian
Roderick Nash has called the movement the "Gospel of Ecology" because
of the religiouslike fervor of ecology advocates. Donald Flemming has
called it the "new conservation," while many ecologists in a special 1964

---

[37] Scott, *City Planning*, pp. 307–309; Edward Basset, *The Master Plan* (New York:
Russell Sage Foundation, 1937), quoted in Scott, *City Planning*, p. 350. The first comprehensive land-use plan in a California county was made in Marin County, a coastal county, in
the 1930s (Scott, *City Planning*, p. 349).

[38] Scott, *City Planning*, p. 350; Rachel Carson, *Silent Spring*, pp. 45–48, 125–26,
135–36, 140, 141, 146–48, 150–51, 187–98, 239.

edition of *Bioscience* referred to it as the "new ecology." Beneath the diversity of titles, the message to the American public was the same; ignorant human interference with the necessary operations of the biosphere would bring life on earth to an apocalyptic end. *Ecology*, through the printed and visual medias, became a household word, and the emergence of a national estuarine preservation ideal was part of this greater mass concern for environmental integrity. The aquatic and oceanic aspects of this grass-roots revolt were popularized by Rachel Carson.[39]

It is to her credit that Carson, who stood apart from the widely accepted national sentiments for water and wildlife, was able to influence a generation of confident affluence. In contrast to popular culture and artistic and literary traditions, Carson refused to share even the common attitude toward one of the sea's most feared villains. Once, aboard the research vessel "Albatross" as part of a Fish and Wildlife survey of the George's Bank, she wrote, "There was something very beautiful about those sharks to me—and when some of the men got out rifles and killed them for 'sport' it really hurt me." Carson might understand the scientific principles on which Leopold based his sport-hunting ethics, but she clearly had no sentiment for such endeavors when they were removed from the necessity of subsistence. This sensitivity to life led her to observe that "by every act that glorifies or even tolerates such moronic delight in killing we set back the progress of humanity." She dedicated *Silent Spring* to Albert Schweitzer, the Alsatian emigré-physician of equatorial Africa who popularized the ideal of "reverence for life." Rachel Carson's respect for nonhuman evolution was basic to her concern for the coastal environment. It was exemplified by a habit she had concerning specimens collected for sketching for *The Edge of the Sea*. After her artist-collaborator had finished with them, "Rachel would put them back in the bucket and return them to their natural places on the beach."[40]

When she received the National Book Award for *The Sea Around Us*, Carson told the gathering, "Despite our own utter dependence on the earth this same earth and sea have no need of us."[41] As a critic of anthropocentric ethics, upon which regional planning rested, Carson revealed an

---

[39] Nash, *American Environment*, pp. 225–27, 238–47; Eugene Odum, "The New Ecology," *Bioscience* 14 (July, 1964): 14; Fleming, "Roots of the New Conservation," pp. 1–7.

[40] Brooks, *House of Life*, pp. 8, 117, 129; Sterling, *Sea and Earth*, p. 122.

[41] Brooks, *House of Life*, p. 129.

understanding of estuaries' role in the productivity of the sea. As the meeting place of land and sea, the shores of America served as a unique testament to life's evolutionary past and a recurrent reminder of human insignificance, Carson felt. In an article for *Holiday* magazine in 1958, she developed this theme—"the ocean has nothing to do with humanity." Carson disapproved of the fact that only 6.5 percent of the Gulf and Atlantic coasts was in federal or state hands and reminded the nation of a 1935 Park Service survey that had recommended the acquisition of 437 miles of seashore, to increase the amount of publicly owned shores to 15 percent.[42]

Providing the ideological foundation for coastal protection, she recommended that "somewhere we should know what was Nature's way; we should know what the earth would have been had man not interfered. . . ." She helped extend wilderness advocacy from wild rivers and forests to the equally wild but scarcer coast. In this devotion to the preservation of untamed ocean coasts, Rachel Carson reaffirmed the human psychic need to find its identity in relationship to surrounding land and water. She recognized the inseparability of land and life, especially of wetlands and wildlife.[43]

Carson's estuarine-protection imperative was rooted in this realization that coastal vistas lend a sense of geologic history to human endeavor. Her writings demonstrated this invaluable role of seashores in the human imagination. Anyone standing "at the edge of the sea," she felt, could sense in "the ebb and flood of the tides" or in the "breath of mist moving over a great salt marsh" the "knowledge of things that are nearly eternal." She described the "flight of shorebirds" and the running of "the young shad to the sea" as patterns repeated "for untold thousands of years" prior to human evolution. Carson's concept of coastal preservation began with a sensitivity to prehistory and encompassed psychic and utilitarian motives.[44]

Carson used economic arguments for protecting recreation and fisheries from industry. As she confided to Dr. William Beebe in a letter dated April 6, 1948, "I am much impressed by man's dependence upon the ocean, directly and in a thousand ways unsuspected by most people." She concluded, "We will become even more dependent on the ocean as we destroy the land."[45]

[42] Carson, "Our Ever-Changing Shore," reprinted in Brooks, *House of Life*, pp. 216–28, especially pp. 219, 225.
[43] Ibid., p. 226.
[44] Brooks, *House of Life*, p. 32.
[45] Ibid., p. 110.

It had been Carson's life-long desire to visit one of the coast's most awesome and rare habitats, the redwood forests of the Pacific shore. She did so in a wheelchair in 1963, attended by David Brower, then executive director of the Sierra Club, on a tour of Muir Woods National Monument. She had known for some time before the writing of *Silent Spring* that she was dying of cancer. After a heart attack in the early 1960s, she had written a friend concerning the irony of the fame and influence she had won with the publication of her last book. "I keep thinking—if only I could have reached this point ten years ago," Carson explained; "now, when there is an opportunity to do so much my body falters and I know there is little time left." [46]

For Rachel Carson the inevitable end came on April 14, 1964, at her home in Silver Spring, Maryland. Concerning death, she had written her closest friend, Dorothy Freeman, the previous September, "When any living thing has come to the end of . . . that intangible cycle . . . , it is a natural and not unhappy thing that a life comes to its end." The entire note was read at her funeral service on April 19. Carson left a revived legacy of coastal appreciation that numerous authors continued to popularize. [47]

Polly Redford in a 1967 article for the *Atlantic Monthly*, entitled "Vanishing Tidelands," rekindled the concern Carson had aroused for estuary preservation. She explained the need for public support of the National Estuary Protection Bill under consideration by Congress. Two years later John and Mildred Teal, once associated with Odum's Sapelo Island research team, published the first book to deal with East Coast estuaries in a popular and comprehensive manner, *Life and Death of the Salt Marsh*. These advocates of tidal-marsh protection led the national political debate over environmental quality, scenic beauty, and economic development to focus on the least appreciated habitats of the coastal regions. They accomplished what death had prevented Rachel Carson from doing. [48]

The combined efforts of these naturalists contributed to the articulation of an estuarine preservation ideal against which political efforts to conserve coastal resources and amenities could be measured. The rationale they provided was multifaceted. Not only was the estuary "a keystone for marine life because of its position in the food web of the sea," as one biolo-

---

[46] Ibid., pp. 314, 321, 323.

[47] Ibid., pp. 326, 327.

[48] Polly Redford, "Vanishing Tidelands," *Atlantic Monthly*, June, 1967, pp. 75–83; John Teal and Mildred Teal, *Life and Death of the Salt Marsh*, pp. 179–263; Peggy Weyburn, *The Edge of Life*.

gist, Bruce Wallace, explained. In their pristine condition, marshes also absorbed biodegradable wastes in such volumes that their value for sanitation purposes was estimated at upwards of $50,000 per acre in money required to build comparable sewage-treatment plants. Another economic rationale was the services coastal swamps provided as low-cost flood-control and water-recharge areas. As Wallace admonished, tidal salt marshes were as essential to the nation's food supply and healthy economy as the Midwest's fertile farms. The long tradition of discounting wetlands was being rejected.[49]

[49] Bruce Wallace, *Essays in Social Biology*. Vol. 1: *People, Their Needs, Environment and Ecology*, pp. 244–45; Hanie, *Guale*, pp. 20–28; issue entitled "America's Changing Environment," *Daedalus* 96 (Fall, 1967): 1003–19, 1142–57, 1184–91.

8

Politics and the Preservation of Estuaries

> The race between education and erosion, between wisdom and waste
> has not run its course. . . . Each generation must deal anew with the
> raiders, with the scramble to use public resources for private profit,
> and the tendency to prefer short-run profits to long-run necessities.
> The nation's battle to preserve the common estate is far from
> won. . . . The crisis may be quiet, but it is urgent. . . . We must
> expand the concept of conservation to meet the imperious problems
> of the new age.
>
> —John Fitzgerald Kennedy

THE pejorative implications of the words morass, slough, muck, and miasma still associated with wetlands need no comment. These terms are common synonyms for obstruction, nuisance, and disease. Yet as part of the
national "battle to preserve the common estate," coastal marshes and tidal
flats became the focus of a major drive for protection of natural resources
during the late 1950s.

As President Kennedy suggested, the meaning of conservation was in
a state of flux, and the concept reappeared constantly in the popular literature after the war as an antonym for development. This differed from the
earlier notion of conservation offered by Gifford Pinchot, who liked to believe that "the first principle of conservation is development." However,
he too recognized an internal ambiguity in the term, noting that conservation included "development, preservation, the common good." The influence of the new science of ecology redefined conservation after 1945 and
differentiated among the related concepts of conservation, development,
and preservation.[1]

Development differs from conservation, in that the motivation for development is to serve the good of the whole by rewarding individual risk
while conservation attempts to promote the general welfare by denying the

[1] Gifford Pinchot, *The Fight for Conservation*, pp. 42–43, 48–49.

individual the right to harm common properties when taking that risk. Preservation, a third related concept, seeks to encourage individual and common support for the protection of cultural or natural artifacts of outstanding biological, technological, or social importance. Often these preserved places or things represent important traditions, folkways, or landscapes having symbolic value that adds to the cohesion and stability of a human group over time.[2]

Preservation protects historic buildings, scenic vistas, or ecologically sensitive coastal areas from development. Cape May, New Jersey; Key West and Saint Augustine, Florida; Galveston, Texas; and Savannah, Georgia, offer examples of coastal historic districts encouraged by local private interests, state conservancies, and federal legislation. Such cooperative effort is part of the same tradition in which private owners donated the scenic and cultural resources of Mount Desert Island, Maine, to the National Park Service in 1916. Although related to preservation of cultural resources, protection of wetlands for their bioeconomic values differs substantially from traditional utilitarian and aesthetic reasons for preservation of wildlife habitats and scenic beauty.[3]

In place of Progressive definitions of conservation, in the postwar period, at both grass-roots and bureaucratic levels, an ecocentric understanding evolved, which viewed wise use (conservation) and protection (preservation) as mutually supportive. Undergirded by the scientific researches of the new ecology, this coalition of preservationists and conservationists produced a range of state and federal legislation that embodied what Lyndon Baines Johnson called the new conservation. In an address on February 8, 1965, he characterized the movement as concerned "not with nature alone, but with the total relation between man and the world around him."[4]

Many obstacles arose to limit the influence and widespread acceptance of conservation's new meaning—first and foremost, the strategic position of estuaries in the nation's demographic expansion. Various esti-

[2]Charles H. Hosmer, *The Presence of the Past: A History of the Preservation Movement*, pp. 41–100; Hiroshi Daifuku, "The Conservation of Cultural Property," in *The Fitness of Man's Environment*, ed. Robert M. Adams et al., Smithsonian Annual No. 2 (Washington, D.C.: Smithsonian Institution Press, 1968), pp. 191–205.

[3]Stewart Udall, *The Quiet Crisis*, p. 162; Hans J. Huth, *Nature and the American: Three Centuries of Changing Attitudes*, pp. 116–19.

[4]U.S. Department of the Interior, *Yearbook 1965* (Washington, D.C.: Government Printing Office, 1966), pp. 3–45.

mates placed the bulk of the country's population within fifty miles of a lake, estuarine, or outer coastal shoreline. The majority of energy-generating facilities, pulp mills, pesticide factories, phosphate mines, chemical plants, and oil refineries competed with residents for space along the coasts.[5] Furthermore, postwar affluence not only added to the stream of new pollutants flowing into coastal waters but also increased the demand for seaside recreation facilities, fresh water, and open space.

Although the average density of the United States is eighteen times less than that of Holland, where in 1969 there were estimated to be 816 persons per square mile, the figures give an incomplete picture. Since the 1930s it has been popular to characterize the area of new urban and suburban growth in the nation as the Sun Belt, suggesting the national demographic shift to California, Texas, Arizona, and Florida. The term, though, is misleading. Growth since the end of the depression has occurred most dramatically in the coastal zone. California, Texas, and Florida, which together hold more than half the nation's coastal acreage, have led the national average in growth.[6]

A book by Anne W. Simon concerning the coastal crisis in the United States has correctly identified the seashore as "a magnet for people." Simon has explained that 53 percent of the nation's population crowd the shore, including the Great Lakes, and this area has accounted for 90 percent of our postwar growth. These shorelines include fifteen of the twenty largest standard metropolitan statistical areas in the country. In every case the general density of coastal states is exceeded by the density of their coastal counties. The average density of New York State is 351 persons per square mile, but for coastal counties it is over 5,000 persons per square

---

[5] William A. Niering, "The Dilemma of Coastal Wetlands: Conflict in Local, National and World Priorities," in *The Environmental Crisis*, ed. Harold W. Helfrich, Jr., pp. 143–51; Paul Ehrlich, *The Population Bomb* (New York: Sierra Club/Ballantine, 1968), pp. 47–48, 51–57, 64–67, 127; Paul R. Ehrlich, Anne H. Ehrlich, and John P. Holdren, *Human Ecology* (San Francisco: Freeman, 1973), pp. 52–53, 97–106, 141, 150, 166–70, 190–92, 215–19.

[6] Philip Nobile and John Deedy, eds., *The Complete Ecology Fact Book* (Garden City, N.J.: Doubleday, 1972), pp. 22, 30, 41; Chauncy D. Harris, "The Metropolitan Districts in 1940," *Journal of Geography* 41 (December, 1942): 340–43; Chauncy D. Harris, "Growth of the Larger Cities in the U.S., 1930–1940," *Journal of Geography* 41 (November, 1942): 313–19; *The World Almanac: 1976* (New York: Newspaper Enterprise Association, 1976), pp. 98, 250.

mile. The population of Long Island alone is greater than that of all but seven states and represents well over one-third of New York State.[7]

In southern California's sprawling coastal counties, the average density is far closer to the national average of 100 persons per square mile. But in the mile-wide coastal zone, density increases to 391 persons per square mile, and the figure jumps to 3,980 people per square mile along the actual waterfront. By comparison, the Chesapeake Bay region averages 940 persons per square mile of shore. Recent census returns suggest that the coastal and lake shores of the nation have undergone a population explosion from which they may not soon recover.[8]

Coastal problems—and planning for their solution—are compounded by the per capita demands of the national life-style and federally subsidized improvement of the standard of living. The demand for seaside recreation, for instance, encounters certain problems and raises certain others. The creation of national seashores to be enjoyed by coastal tourists invariably irritates local residents and decreases local property-tax revenues. Local citizens may therefore oppose efforts to open the littoral to a larger public. Shore access by foot or bicycle is often limited by parking lots, when not put altogether off-limits by private ownership. Furthermore, consumer preferences and public policy choices have eliminated the bus, train, or ferry ride an earlier generation could take to get to the beach. Reliance on automobiles and highways instead has seriously curtailed the numbers of people who can reach the shores. As Lewis Mumford has pointed out, railroads can transport 40,000 to 60,000 people per hour along a single narrow corridor, whereas the largest expressways are built to handle only 4,000 to 6,000 cars in the same period.[9] On the other hand, limitations on coastal access may be a blessing in disguise. For instance, the U.S. Army

[7] Anne W. Simon, *The Thin Edge: Coast and Man in Crisis*, pp. 17–18; Jerome Pickard, "U.S. Metropolitan Growth and Expansion, 1970–2000, with Population Projections," in *Commission on Population Growth and the American Future: Research Reports*, vol. 5, ed. Sara Mills Mazie (Washington, D.C.: Government Printing Office, 1972), pp. 130–59; *World Almanac: 1976*, pp. 204–209; House, Committee on Merchant Marine and Fisheries, *Estuarine and Wetlands Legislation. Hearing Before the Subcommittee on Fisheries and Wildlife Conservation*, 89th Cong., 2d sess., June 16, 22–23, 1966, pp. 20–51.

[8] Senate, *The National Estuarine Pollution Study. Report of the Secretary of the Interior*, Sen. Doc. 91-58, 91st Cong., 2d sess., March 25, 1970, pp. 144–45.

[9] Mumford, quoted in Garrett DeBell, ed., *The Environmental Handbook* (New York: Sierra Club/Ballantine, 1970), pp. 66–76, 182–96; "Rim of the Sea Becoming Crowded and Dirty," in *Man's Control of the Environment*, ed. Jack Nease, p. 26; Wesley Marx, "How Not to Kill the Ocean," *Audubon Magazine*, July, 1969, pp. 27–35.

Chief of Engineers remarked on September 11, 1969, "Seekers of outdoor recreation are . . . creating 50-mile long weekend traffic jams reaching all the way from Boston to Cape Cod, crowding the rivers and waterways with their motorboats . . . and demanding service and accommodation." The ecological integrity of the coasts is threatened by such heavy use, and the local infrastructure may be overburdened by it. Taxes extracted from coastal areas must, at least, cover the costs required to maintain healthy estuaries, lest falling recreation fees and tourist taxes further deplete coastal economies.[10]

Other needs than recreation contribute to undermining the health of seashores, whether commonly or privately owned. Rising demands for cement for shelter and paving pose one problem. The estuary may seem a strange place for mining operations, but the shell, sand, gravel, clay, and lime used in cement have been hewn from our coasts as well as from interior wetlands or rivers. In 1882, eleven years after portland cement was invented in Allentown, Pennsylvania, the nation produced 85,000 barrels per year. By 1963, the quantity had increased over 4,000 times to 361 million barrels per year.[11] Similarly, the 500 percent increase in the demand for electricity between 1950 and 1971 required more utilization of coastal resources. Tied closely to the demand for new homes, the need for electrical energy was expected to increase from the 845 billion kilowatt hours per year in 1960 to over 2 trillion kilowatt hours in the 1980s. A large number of the power plants built to supply the necessary electrical energy in the 1960s were in the coastal zone. These posed serious threats to the littoral through their future land-use needs and through discharge of the plants' heated waste water.[12]

Water shortages and deterioration of water quality have presented the most persistent threats to coastal ecology. At the turn of the last century the nation used 40 billion gallons of water per day. Although figures vary, by 1965 an estimated ninefold increase had occurred, representing the use of 1,900 gallons per person every day. While it has been estimated that the average American requires about 150 gallons per day, including 1 gallon to drink, most water needs are indirect. The wheat crop alone requires 300

[10] "Rim of the Sea," p. 26.
[11] Senate, *Estuarine Pollution Study*, pp. 132–33; Hans H. Landsburg, *Natural Resources for U.S. Growth*, p. 98; Peter Flawn, *Mineral Resources* (Chicago: Rand McNally, 1966), pp. 47, 128–35.
[12] Landsburg, *Natural Resources*, pp. 8–9, 98.

gallons per day, and when the human food requirement is included, the per capita daily water requirement jumps to 2,500 gallons. Synthetic chemicals are a primary water-dependent industry, consuming between 100,000 and 200,000 gallons of water per ton of rayon. Compared with this, the 50,000 gallons per ton of steel or 39,000 gallons per ton of finished paper appear miniscule. Permanent water shortages have been projected for the nation by the year 2000. Before that, the saltwater intrusion throughout most of the nation's estuaries may permanently alter the biology of the seashore.[13]

By the 1960s the effects of urbanization and industrialization were being recognized by a growing number of Americans. The ecological awareness fostered by Aldo Leopold, Rachel Carson, and others was ripe for translation into public policy. Individuals at the local level and in all branches of the national government began to act. Stewart Udall, named secretary of the interior by President John Kennedy in 1961, was one of the important federal officials who popularized both the need for environmental activism on behalf of commonly owned natural resources and the necessity of new sources of funding for that protection. In his widely influential book *The Quiet Crisis* he echoed the warning he gave as secretary of the interior: "As our land base shrinks it is inevitable that incompatible plans involving factories, mines, fish, dams, parks, highways and wildlife, and other uses and values will increasingly collide." He urged citizens to realize that this "quiet crisis demands a rethinking of land attitudes. . . ." "True conservation," he insisted, "is a thing of the mind—an ideal of men who cherish their past and believe in the future." The growing urgency of translating such ideals into viable political options was apparent to the Arizona native. Udall placed the influence of the new ecology on the Kennedy-Johnson environmental program in the earlier tradition of the two Roosevelts, referring to most recent efflorescence as the "third wave of conservation." Nowhere did the conflict among competing uses for the land become more evident than along the coast, and Udall tried to set priorities for resolving it. Under his direction, six national seashores were

[13] Flawn, *Mineral Resources*, pp. 7, 9, 45, 237; Bernard Frank and Anthony Netboy, *Water, Land, and People* (New York: Alfred Knopf, 1950), pp. 9–11; John Bardach, *Downstream: A Natural History of the River* (New York: Harper and Row, 1964), pp. 230–32; *Congress and the Nation* (Washington, D.C.: Congressional Quarterly Service, 1968), II, 495; Robert Theobald, *Habit and Habitat*, p. 85; Udall, *Quiet Crisis*, p. 191.

added to the public domain, whereas previously only Cape Hatteras and Mount Desert Island had been protected.[14]

He was aided by powerful senators and congressmen with environmental constituencies, such as Senators Warren Magnuson of Washington, Hubert Humphrey of Minnesota, Robert Kennedy of New York, and Edward Kennedy of Massachusetts, and Congressmen John Dingell of Michigan and Herbert Tenzer of New York. Their concern and interest had grown during the 1950s to include wilderness preservation, water-pollution control, support for the National Park Service, and maintenance of public access to the shoreline.[15]

Support for ecological measures began to appear in the Supreme Court, too, especially from Justice William O. Douglas. Even before the New Frontier launched the "third wave of conservation," Douglas had influenced the legal foundations for sustaining environmental legislation. Early in 1941, Douglas was instrumental in interpreting the constitutional authority of the national government to control interstate commerce and navigable waterways to include greater federal control over nonnavigable portions of a river. Douglas not only broadened the concept of navigation to include flood control, but also held that just because "ends other than flood control will also be served . . . does not invalidate the exercise of authority conferred on Congress." [16]

Thirteen years later, in 1954, in an equally significant ruling, this time concerning land use for redevelopment, Justice Douglas, writing for the majority, held that under the exercise of their police power the states could achieve aesthetic ends in planning and zoning. Although the decision involved Congress's powers over the District of Columbia, Douglas held that these powers were vested in all legislatures. "If those who govern the District of Columbia decide the Nation's Capital should be beautiful as well as

[14] Udall, *Quiet Crisis*, pp. 167, 194, 195, 196, 199, 200; Harold Gilliam, "Udall: Chief Architect of the New Conservation," *This World Magazine, San Francisco Sunday Chronicle and Examiner*, January 19, 1969; "Rim of the Sea," p. 27.

[15] Richard A. Cooley and Geoffrey Wandesforde-Smith, eds., *Congress and the Environment* (Seattle: University of Washington Press, 1970), pp. xii–xv; Senate, Committee on Commerce, *Estuaries and Their Natural Resources. Hearing on H.R. 25 and S. 695*, 90th Cong., 2d sess., June 4, 1968, pp. 2–27.

[16] Alfred H. Kelly and Winfred A. Harbison, *The American Constitution: Its Origins and Development*, pp. 778–80; Mel Scott, *American City Planning*, p. 492.

sanitary," he wrote, "there is nothing in the Fifth Amendment that stands in the way." [17]

During the environmental movement of the sixties, Douglas called for a "wilderness bill of rights" that would guarantee, among other things, the preservation of wetlands, protection of fisheries from obstructing dams, and provision of wildlife sanctuaries along coastal areas. He referred to the Dismal Swamp (coastal Virginia and North Carolina) as holding an educational potential to offer "graduate courses in biology, botany and natural wonders." Once, along the coast of Avala Point, Washington, he exclaimed, "I always leave this primitive beach reluctantly," because "the music of the oceanfront seems to establish a rhythm in man. For hours, even days afterward I can almost hear the booming of the tides on the headlands. . . ." [18]

With such sentiments in the federal high court shared by Congress and the executive branches, the national political machinery was uniquely ready to make significant progress in environmental policy. Legislative formulation of a national estuarine-preservation ideal was, in part, the federal response to scientific advice to government. To an even greater degree it grew out of grass-roots support from the nation's various quarters. Local groups, including middle-class homeowners and intellectuals, came together in opposition to the varieties of development that threatened the biological integrity, scenic beauty, or economic value of coastal wetlands on the Atlantic, Gulf, and Pacific shores.

Estuaries and lagoons were the focus of three such debates that began in San Francisco, New York, and Massachusetts and alerted congressmen to the necessity of national guidelines concerning coastal land- and water-use decisions. Unlike the Delaware or Chesapeake Bay estuaries, the regions at question rested entirely within the jurisdictions of their respective states.

Regional planning in northern California, like that in the Delaware Valley or New York Bay areas, has existed more as sentiment than as hard, realistic agencies with coercive power. From 1906, when the Burnham Plan advocated comprehensive city planning for San Francisco, steps toward areawide, long-range plans were ineffective. The existence of nine counties hindered agreement, while east-bay cities frequently resisted the

[17]Scott, *City Planning*, p. 492.

[18]William O. Douglas, *A Wilderness Bill of Rights* (Boston: Little, Brown, 1965), pp. 12–13, 32, 80–83, 126–27, 128–29, 144–46; William O. Douglas, *My Wilderness: The Pacific Northwest* (Garden City, N.J.: Doubleday, 1960), pp. 35–37, 49.

desire of San Francisco to dominate such plans. Finally in 1957 a loose confederation, more a forum for research and discussion of common problems than a planning authority, was created and was designated the Association of Bay Area Governments (ABAG). Much as in the impetus for Penjerdel in the Delaware Valley, local governments in San Francisco were primarily concerned over air pollution and highway routes. Then during 1959 two events sparked heated controversies from which a grass-roots environmental constituency emerged among Bay Area residents.[19]

The Army Corps of Engineers published a pamphlet in 1959 projecting the filling of the bay's wetlands until only 187 of 435 square miles would have existed by the year 2000. The margins of the estuary were to be used for housing, transportation, and industry, and since they were publicly owned the cost of their transformation would include merely the value of labor and technology, while the benefits would add considerably to the local tax base. Fifty miles north of the Golden Gate, future demands of a different kind centered on the sleepy fishing village of Bodega Bay. During the late 1950s, the Pacific Gas and Electric Company had studied a rocky promontory as the future location of its first nuclear-power plant. Abundant supplies of cold water assured the necessary cooling of the reactor, and the remoteness of the area guaranteed that no large populations would be affected. While these proposals appeared necessary for the future growth and prosperity of the greater bay region, subsequent events, the callous disregard of public sentiments by a few corporate officials, and an alliance of local preservationist groups upset the plans. Opposition developed on the grounds that the projects were hazardous and were economic threats to the region's inhabitants, resources, and amenities.

Although fishermen and citizens of Sonoma County rallied to thwart the Bodega Bay plant, it was the discovery by scientists at the University of California and in the U.S. Geological Survey that the nuclear facility rested on the San Andreas Fault that influenced the Division of Reactor Licensing in 1964 to declare Bodega Bay unsuitable for an atomic power plant. The fury with which local groups attacked the indiscriminate filling of San Francisco Bay had far greater implications for comprehensive regional

---

[19] Mitchell Postel, "Vigil of the Golden Gate: The Environmental History of San Francisco Bay since 1850" (Master's thesis, University of California, Santa Barbara, 1977), pp. 32–100; Scott, *City Planning*, pp. 63–65, 383–85; Wesley Marx, *The Frail Ocean*, pp. 155–69; Harold Gilliam, "The Fallacy of Single-Purpose Planning," *Daedalus* 96 (Fall, 1967): 1142–57.

planning than the Bodega Bay incident. Yet both of these issues—the siting of coastal power plants and the filling of tidelands for short-term economic reasons—revealed, in the words of an influential local naturalist, Harold Gilliam, "that there exists in the State of California no agency to protect the people's interest in maintaining scenic open space." [20]

Among the several grass-roots organizations that emerged to force the creation of an area-wide agency to forestall filling and adequately plan for future development were the Save the San Francisco Bay Association and the Citizens for Regional Recreation and Parks. Both were assisted by local membership drives and guided by people associated with the University of California. Faculty wives from Berkeley led the save-the-bay group, which stressed aesthetic values of this picturesque landscape. The other group was presided over by Mel Scott, who had been affiliated with Berkeley's Institute for Governmental Affairs. His studies emphasized the economic losses from destruction of the bay's amenities, fisheries, or water quality, and the organization circulated his findings in a monthly newsletter. At its peak the newsletter reached a circulation of over sixteen thousand in 1968, while the Save-the-Bay membership swelled to nearly nine thousand in the summer of 1965. Along with the Sierra Club, the League of Women Voters, and the Federation of Western Outdoor Clubs, these volunteers exposed ABAG's powerlessness to protect estuarine wetlands from destruction and the complicity of the Army Corps of Engineers in permitting dredge-and-fill operations. As the only authority with regional competence, the Corps was technically in violation of the Fish and Wildlife Coordination Act of 1934. [21]

Citizens groups pressured the governor to call a special session of the legislature in 1964 to create a state commission that would counter the piecemeal, single-purpose, laissez-faire development of the bay's wetlands by imposing a moratorium on fill and preparing a study of the region with long-range planning in mind. In June, 1965, the Bay Conservation and Development Commission (BCDC) was created, largely through the leadership of State Senator Eugene McAteer. The moratorium on bay fill took effect in September, 1965, and the citizens' groups were instrumental in

[20] Michael J. Barbour, Robert B. Craig, Frank R. Drysdale, and Michael T. Ghiselin, *Coastal Ecology of Bodega Head* (Berkeley: University of California Press, 1973), pp. 213–39; Gilliam, "Fallacy of Single-Purpose Planning," pp. 1143–46, 1151–55; Polly Redford, "Vanishing Tidelands," *Atlantic Monthly*, June, 1967, pp. 76–77, 83.

[21] Postel, "Vigil of the Golden Gate," pp. 30–55.

convincing the legislature to make the BCDC a permanent institution in 1969. Although attempts to secure further environmental measures from the state legislature failed repeatedly, the BCDC did become the model for the California Coastal Commission, which was created by a citizens' ballot initiative (Proposition 20) in 1972.[22]

In January, 1970, two tankers collided in San Francisco Bay and dumped more oil into the estuary than the blowout of Union Oil Company's drilling rig had spewed into the Santa Barbara Channel during the previous winter. Both of these disastrous consequences of maritime oil exploration and transport pushed California toward a reexamination of its coastal regulations. Bills for state-wide coastal planning introduced in 1970 opened a two-year debate between coastal-preservation coalitions and real-estate developers, which ended in the passage of Proposition 20. In mandating comprehensive planning of coastal resources, the ballot initiative created county, regional, and state coastal commissions that virtually deprived local real-estate interests of absolute control over coastal development. In October, 1969, the *Los Angeles Times* had editorialized that such legislation was needed to aid "in establishing a reasonable policy in the protection of estuaries, such as upper Newport Bay."[23]

The politics of the California seashore were not unique; similar conditions had engendered studies and agencies in the densely populated northeast corridor from Boston to the Chesapeake Bay. In fact, problems concerning the multiple uses of coastal resources were originally addressed by New York primarily because of the astounding growth of Long Island, and early agitation for a national estuary-protection policy can be traced to the island's suburban counties.[24]

In 1952 the Conservation Department of the State of New York began the first of numerous surveys of Long Island's wetland acreage. Two years later the federal Bureau of Sport Fisheries and Wildlife did a study of Long Island wetlands and the tidal marshes from Maine to Delaware that supplemented state efforts. That same year interstate controversies over water allocation brought about the creation of the Delaware River Basin Commis-

[22] Ibid.; Gilliam, "Fallacy of Single-Purpose Planning," pp. 1152–54; Stanley Scott, "Coastal Planning in California: A Progress Report," *Bulletin of the Institute for Governmental Affairs* 19 (June–August, 1978): 1; Simon, *Thin Edge*, pp. 150–56.

[23] "Big Oily Mess in the Bay after 2 Tankers Collide," *San Francisco Chronicle*, January 19, 1971, p. 1; *Los Angeles Times*, October 12, 1969 (editorial); Robert G. Healy, *Land Use and the States*, pp. 64–102.

[24] House, *Estuarine and Wetlands Legislation*, pp. 20–51.

sion designed to research, recommend, and execute policies to protect and manage the water resources of this multistate region. As they had on the Pacific Coast, land, water, and energy needs of urban populations forced state and later federal action despite vociferous protests from local townships and some private industries.[25]

The controversies that led to the passage of the National Estuary Protection Act in 1968 began as a crisis over coastal wetlands in Nassau and Suffolk counties, on Long Island, in New York. Under the state law providing for acquisition of tidelands for preservation and conservation, the Town of Hempstead in 1963 set aside 2,500 acres of tidal marsh on the Great South Bay. The purchase grew out of a recommendation made by state and federal officials. The U.S. Fish and Wildlife Service in conjunction with the state Conservation Department had conducted surveys of Oyster Bay and Hempstead townships' wetlands between 1957 and 1961. Their 1961 report, incorporating the research of Odum, Gunter, and others, suggested a management plan for the town-owned wetlands, amounting to 15,500 acres. In June of 1965, a larger survey by the Fish and Wildlife Service revealed that 29 percent of Long Island's wetlands had disappeared between 1954 and 1964 and 88 percent of what remained was "vulnerable to destruction."[26]

Large numbers of people had moved to the island to escape the deteriorating quality of New York City's environment, tripling Nassau County's population from 1936 to 1964. The county, though, was repeating some of the city's worst mistakes in the management of its water-dependent resources. At the turn of the century, there had been forty square miles of marshland remaining in the five buroughs of New York City. In 1975 less than six square miles of tidal marshes remained along the city's extensive waterfront.[27]

The studies of Long Island during the 1950s and 1960s revealed more than just wetland destruction from dredging and filling operations. For twenty years state and local public-health authorities in New York had relied on DDT to control mosquito-borne diseases. In 1967 a report published by George M. Woodwell concerning studies of Long Island estuaries concluded that toxic levels of DDT had been concentrated by filter-feeding organisms and threatened the future of the island's bird life. As

[25] Ibid., pp. 22–23, 127.
[26] Ibid., pp. 25–26.
[27] Ibid., pp. 113–15; David Allen Gates, *Seasons of the Salt Marsh*, p. 87.

Rachel Carson had warned in *Silent Spring*, through a process of biological magnification DDT residues washed into the estuaries from non-point sources of pollution—that is, from no single locale or source—were concentrated ten million times, from the water content of three parts per trillion to some bird tissue content of twenty-five parts per million. Later studies proved that magnification of pollution from pesticides, heavy metals, and PCB in terrestrial and marine food webs had so weakened the shells of birds that several estuarine bird species—including brown pelicans, clapper rails, perigrine falcons, bermuda petrels, golden eagles, and bald eagles—were facing extinction.[28]

If reclamation, pesticide pollution, and non-point sources of coastal contamination threatened the biological integrity and productivity of nearshore waters, dumping practices posed a more immediate threat to New York–area fishes. One form of dumping was heated-water discharge from Consolidated Edison's Storm King generating station on the Hudson River; the second was disposal of solid waste in the Hudson submarine canyon on the Continental Shelf. In 1963 a nuclear-power plant was opened at Indian Point, on the tidal waters of the Hudson River. Water used for cooling the facility was returned to the Hudson River seven degrees warmer than when it had been removed. In testimony before Congressman John Dingell's Committee on Fisheries and Commerce, the Atomic Energy Commission's director of regulation, Harold Price, maintained that under existing law he could not halt the construction of a nuclear plant because of thermal wastewater discharges, because conventional facilities were not controlled in such a manner.[29]

Robert Boyle, then conservation editor of *Sports Illustrated*, had objected to the facility's having been built beside the spawning grounds of an important sport fish, the striped bass. "You do not build a power plant on top of the spawning bed of an irreplaceable fish resource," he argued. Striped bass spend much of their adult lives in the sea but are caught in estuaries during their migrations to spawn in the fresher reaches of the tide-

[28] George M. Woodwell, "Toxic Substances and Ecological Cycles," *Scientific American*, March, 1967, pp. 24–31; George M. Woodwell, Charles F. Wurster, Jr., and Peter A. Isaacson, "DDT Residues in an East Coast Estuary: A Case of Biological Concentration of a Persistent Insecticide," *Science*, 1967, pp. 821–23; George M. Woodwell, Paul P. Craig, and Horton A. Johnson, "DDT Residues in the Biosphere: Where Does It Go?" *Science*, 1971, pp. 1101–1107.

[29] Marx, *Frail Ocean*, pp. 90, 91, 95–97; Arnold Reitze, ed., *Environmental Planning: Law of Land and Resources*, §14, pp. 21–33.

waters. Unlike salmon or trout, which swim the entire length of a river, striped bass spawn in the null zone or mixing place between fresh and salt waters. Like salmon, shad, herring, and alewives, they feed in estuarine waters during their maturation and are also caught by fishermen above the Continental Shelf. A report in 1966 revealed that over 60 percent, by weight, of the fishery catch in the United States was in some way estuarine-dependent. Thus in the mid-1960s, the future of some 90,000 commercial fishermen depended on the food-producing capacities of tidal marshes and the water quality of estuaries.[30]

The second incident of alerting New York residents to the insufficient power of local politicians to control development along estuaries was the 1963 approval by the Army Corps of Engineers of the dredging of a previously established shellfish sanctuary at the request of the Hempstead town board. Incensed citizens formed the Hempstead Town Lands Resources Council, which swelled to more than twenty-five thousand members. Beginning in 1965, Congressman Herbert Tenzer of Long Island introduced federal legislation to aid Hempstead's half-hearted efforts to protect its commonly owned wetlands from land developers, mining, and marinas in the adjacent tidal marshes. Preservationists argued that wetland protection was beyond the control of local municipalities. With state aid on September 23, 1965, the Town of Hempstead set aside an additional 10,000 acres of tidal wetlands. In introducing a resolution for the creation of a Long Island National Wetland Area that year, Tenzer explained, "Long Island . . . property owners without special interests, want these 16,000 acres to be protected in perpetuity." Tenzer wanted to secure a national policy to encourage the protection of these productive fishery and marsh areas because of the capricious character of local control of commonly held tidelands. Tenzer argued that the town's initial "dedication did not reflect a change in policy, but rather an appeasement of conservationists, . . . who are fighting bitterly to prevent major bay bottom losses in other sections of Hempstead's valuable wetlands." [31]

By the time legislation for the protection of estuaries was introduced in Congress during 1966, there was a long and continuously expanding in-

<hr />

[30]Marx, *Frail Ocean*, p. 95; Senate, Committee on Commerce, *Estuarine Areas*, 90th Cong., 2d sess., S. Rept. 1419, July 17, 1968, pp. 3–5.

[31]House, *Estuarine and Wetlands Legislation*, pp. 26, 155–62; House, Committee on Merchant Marine and Fisheries, *Estuarine Areas*, 90th Cong., 1st sess., H. Rept. 989, November 28, 1967, pp. 4–9.

terest in comprehensive planning, specifically for water-quality control, fish and wildlife conservation, and wetlands acquisition. Considering the trend toward reliance on comprehensive planning, the increasing environmental concern and the growing scientific evidence about estuarine productivity, the defeat of the 1966 version of the act was a remarkable environmental failure.[32]

Meanwhile, in Massachusetts, a coalition of state conservationists had been lobbying and publicizing the need for state regulation of coastal filling. The Army Corps of Engineers had since 1956 been required to consult with the Fish and Wildlife Service concerning the effects of filling on natural habitats, but in practice strict constructionists had limited the deliberations of the Corps of Engineers in granting permits to consider the effect on navigation only. In 1885 the commonwealth had had fifty-six thousand acres of coastal wetlands, but a three-year water-pollution study revealed that only thirty-one thousand acres remained in 1966. To retard the loss of wetlands, the Jones Act passed by Massachusetts in 1963 required anyone seeking to dredge or fill in coastal wetlands to obtain a permit from the State Department of Natural Resources. The state claimed a constitutional right under its police powers to limit the use of private property if the utilization endangered coastal shell or marine fisheries.[33]

While developers and environmental protectionists battled in the Massachusetts courts and legislative offices, Governor John Volpe announced in 1965 his support for legislation to broaden the permit power of the Department of Natural Resources to consider the adverse effects of reclamation, drainage, or navigational dredging on public health, wildlife protection, and flood control. Supported by the Audubon Society, the Izaac Walton League, Nature Conservancy, and many other groups, a proposal to preserve some forty-five thousand acres of coastland marshes and tidelands passed that year. Massachusetts opted to deny local control over tidal marshes and to invest a state agency with the active power to protect the sensitive portions of the shore zone. Real-estate and development interests brought several court suits alleging that such protective action denied them the right to reap the full marketable potential of their private property.

[32] *Congressional Quarterly Almanac* 20 (1964): 510; *Congressional Quarterly Almanac* 17 (1961): 132.

[33] House, *Estuarine Areas*, pp. 42–44; Charles H. Foster, "Coastal Wetlands Protection Program in Massachusetts," in *Eco-Solutions: A Casebook for the Environmental Crisis*, ed. Barbara Woods, pp. 155–61.

They argued that restricting reclamation is "tantamount to an unlawful taking" of private property without due compensation and is prohibited by the Fifth Amendment, as interpreted in a 1922 decision of Justice Oliver Wendell Holmes: "The general rule is that while the property may be regulated to a certain extent, if that regulation goes too far it will be considered as taking." The position of the State of Massachusetts, on the other hand, reflected another of Holmes's astute remarks concerning natural-resource use: "A river is more than an amenity, . . . it is a treasure. It offers a necessity of life that must be rationed among those who have power over it." [34]

Citizens in northern California, on Long Island, New York, and in Massachusetts have thus not been ignorant of the potential for disaster that lurks beneath the surface of their once pristine wetlands and still lovely beaches. By the late 1950s citizens' groups realized that if they did not act locally to protect the amenities of the coast these aesthetic sites would soon be traded for power plants or suburban sprawl. Efforts were also being made at the federal level, however. In 1968, environmental groups, having learned from earlier attempts at multi-purpose planning, lobbied for the enforcement of ecological protection among federal agencies and supported the passage of the National Estuary Protection Act.

This act grew out of a broader state and federal framework encompassing legislation for clean-water standards, fish and wildlife protection, marshland preservation, estuarine and wetland surveys, and environmental planning.

The oldest enactment to which the intent of the federal protective legislation can be traced is a New Deal measure that attempted to elevate the status of fish and wildlife, which had played only a secondary role in federal conservation policy. The Fish and Wildlife Coordination Act, passed on March 10, 1934, shared many of the goals and shortcomings of subsequent legislation in this field. Passed to deal with an emergency concerning the pollution of the Potomac River, it encouraged, but did not demand interagency cooperation to plan future use of water resources in the river basin. Although clean-water impoundment was uppermost in the minds of legislators, they proposed the establishment of fish-culture stations and migratory-bird resting sites and, where economically feasible, the construction of fishways over dams. However, no money was appropriated for

[34] Reilly, *Use of Land*, pp. 145–50; Ira Michael Heyman, "The Great 'Property Rights' Fallacy," *Cry California* 3 (Fall, 1968): 29–34; Simon, *Thin Edge*, p. 156; Bardach, *Downstream*, p. 258.

these purposes. While the act was largely ineffectual for want of money and enforcement provisions, it was another example of Harold Ickes' protectionist sympathies, which had also promoted the use of fish ladders on the Bonneville Dam. The act would later be extended to estuarine preservation and environmental protection generally. Healthy fish and wildlife are the indicators of the biological integrity of any geographic region, and that, in turn, rests upon the water quality of a specific watershed. In both original water-pollution control provisions and subsequent amendments to the Fish and Widlife Coordination Act, the foundations of water-quality and biological integrity were established in law. Coastal-wetlands protection later rested on these foundations.[35]

The inadequacy of the 1934 act was recognized in Congress's amendments to the Fish and Wildlife Coordination Act in 1946. The new provisions made land acquired for flood control or irrigation available to state or private agents for the purposes of fish and wildlife propagation. Intergovernmental coordination between state, local, and federal interests was mandated, but no penalties were provided for failure to comply. The requirements of this law had been met as long as the Army Corps of Engineers had consulted with the Fish and Wildlife Service (created in 1940), even if the resulting plan was detrimental to the biotic integrity of a river or estuary. The furor of preservationists opposed to the Echo Park Dam in Dinosaur National Monument on the Colorado River in 1956 brought about a further amendment to the Fish and Wildlife Coordination Act in 1958. In addition to consulting with the Department of the Interior concerning the effects of water impoundment on fish and wildlife, both the Department of Agriculture and the Secretary of the Army were to include fish and wildlife conservation measures in the construction and operation of water-resource projects. The amendment gave fish and wildlife preservation equal weight with navigation, flood control, irrigation, and hydroelectric-power generation. Where conflicts in this multipurpose development of water resources arose, Congress held the right to decide priorities. Finally, §662a) of the act required consultation with the Fish and Wildlife Service prior to the issuance of any dredge-and-fill permits for any bodies of navigable water.[36]

---

[35] Harold Ickes, *The Secret Diary of Harold LeClare Ickes* (New York: Simon and Schuster, 1953), I, 182–83, and II, 494; James B. Trefethen, *An American Crusade for Wildlife*, p. 233.

[36] Roderick Nash, ed., *The American Environment*, pp. 183–91; Reitze, *Environmental Planning*, pp. 2/46, 2/53.

Congressional action for water-quality control reflected an incremental and halting effort to abate water pollution. Beginning in 1948, Congress recognized the need to assist local governments in financing sewage-treatment plants. Although the federal government had regarded water pollution as a purely local problem, it thus established the precedent of federal involvement in water-quality control. In 1956 an amendment to that bill increased the loans to 50 million dollars annually, but a 1960 attempt to increase this to 90 million dollars annually was vetoed by President Eisenhower.[37]

Despite increased expenditures and new enforcement provisions in a 1961 amendment, it was not until the Water Quality Act of 1965 that federal authorities really accepted that water pollution was a national problem. The 1965 act strengthened antipollution laws and directed that the states adopt and enforce water-quality standards by 1967 or else have such standards set and enforced by the secretary of the interior. The following year the Clean Waters Restoration Act appropriated 3.5 billion dollars over four years for the construction of treatment facilities. In emphasizing basin-wide antipollution planning, the act provided 3 million dollars for a three-year national study of estuarine water quality.[38]

The outright purchase of wetlands for wildlife habitat preservation is the final arena of legislative preparation for the National Estuary Protection Act. In the early 1950s, the Fish and Wildlife Service conducted inventories of wetlands including coastal marshes. The Department of Agriculture designated 1955 as the "year of wetlands." By the end of the decade, New York State provided loans to local Long Island governments for the purchase of coastal and interior wetlands. This was followed by federal grants for the purchase of glacially formed wetlands in 1961 and again in 1964.[39]

In the area of comprehensive planning, several enforcement provisions of legislation not specifically designed for estuarine protection nonetheless expedited governmental acquisition of or planning for coastal wetlands. Beginning in 1962, Congress authorized federal financial assistance for states to undertake comprehensive water planning and related land-use planning through the creation of river-basin or group river-basin commis-

[37] *Congressional Quarterly Almanac* 22 (1966): 632–35.

[38] Ibid.; *Congress and the Nation*, II, 495.

[39] *The Yearbook of Agriculture: 1955* (Washington, D.C.: Government Printing Office, 1955), pp. 444–49, 478–90, 579–83, 586–96; *Congressional Quarterly Almanac* 17 (1961): 132; *Congressional Quarterly Almanac* 20 (1964): 510.

sions. On the federal level the Water Resources Council was established to oversee and coordinate federal, state, local, and private planning for adequate supplies of water in each particular region. County-wide grants from the federal government to aid in marina and port construction and to aid fisheries and aquiculture were provided in 1965 by the Public Works and Economic Development Act. Aid to urban areas for the acquisition of open space, water development, and land conservation was provided under §204 of the Demonstration Cities and Metropolitan Development Act of 1966.[40]

Local review, comprehensive multipurpose regional planning, and federal grants-in-aid are features shared by all of these legislative attempts to plan for growth. Two further pieces of legislation strengthened the planning and development of coastal resources. The Marine Resources and Engineering Development Act of 1966 created another interdepartmental agency and the Commission on Marine Science, Resources, and Development. One of the lasting contributions of the Commission on Marine Science was its recommendation to create the National Oceanic and Atmospheric Administration (NOAA), which was formed in 1970 and placed under the purview of the Department of Commerce. Enhanced bureaucratic cooperation for comprehensive regional planning was also the intent of the Intergovernmental Cooperation Act of 1968, requiring that the perspectives of all relevant national, state, and local agencies be considered when evaluating federally assisted developments.[41]

It is clear in both President Kennedy's call for regional watershed commissions and the legislative approach to marine resources that the federal government intended that local planning of estuarine development be comprehensive and that all natural resources be included in the planning process. Funding for recreational use of the shores, as well as other portions of the country, was provided by the establishment of the Land and Water Conservation Fund in 1964, although the effectiveness of that legislation has since been crippled. Local, state, and federal review procedures for planned development were made more complex by the passage of the National Environmental Policy Act of 1969, which took effect on January 1, 1970.[42]

[40] Reitze, *Environmental Planning*, §2, pp. 16–26.
[41] Ibid.
[42] Eugene W. Weber, "Comprehensive River Basin Planning: Development of a Concept," *Journal of Soil and Water Conservation* 19 (July–August, 1964): 134–37; *Congressional Quarterly Almanac* 17 (1961): 876; Daniel P. Beard, "Meeting the Costs of a Quality Environment," in *Congress and the Environment*, ed. Cooley and Wandesforde-Smith, pp. 96–111; Scott, *City Planning*, 583–85, 611, 615, 618–19, 623–24.

Taken together, these acts reveal a clear congressional intent to place resource uses and developmental activities on an equal footing. As they pertain to estaurine protection, these various acts required that fish and wildlife, navigation and flood control, extractive industrial use and waste-disposal plans be considered equally. In this sense the formulation of a national estuarine preservation ideal was, in part, the outgrowth of a political desire to coordinate the vast federal and state bureaucracies whose responsibilities overlapped in the coastal zone.

There were several direct precedents for estuarine preservation that should have paved the way for its acceptance by Congress. In 1964 both the Wilderness Act and the Land and Water Conservation Fund Act were passed. In 1963 and again in 1965, Massachusetts adopted legislation to halt the filling of valuable coastal wetlands, and in 1965 the federal government established the fifth national seashore on forty-three hundred acres of Fire Island, the barrier island that protects the south shore of Long Island.

In 1965 congressional proponents of estuarine preservation began to confront the issue directly. In that year Congressman Tenzer introduced his proposal to create the Long Island National Wetland Area. This plan was bolstered by legislation calling for national estuarine protection areas, sponsored by Congressman John Dingell of Michigan and Senator Edward Kennedy of Massachusetts. Support during the June, 1966, hearings came from local conservancy groups, the Audubon Society, and the American Littoral Society. Dingell's plan would have given the secretary of the interior control over dredging and filling permits and acquisition powers to create, after a two-year study, a national system of estuarine preserves.

On October 3, 1966, however, Dingell attempted to bring his administration-backed resolution up for a vote in the House without allowing anyone to amend the draft during floor debate. In order to do this, Dingell was required to get a two-thirds majority of the lower chamber to suspend the rules. His attempt fell three votes short: 212 representatives voted in favor of suspension of the rules.[43]

The defeat of this strong measure for estuarine protection doomed Dingell's 1966 compromise resolution of the original drafts by the two Kennedys and Tenzer. A majority of representatives from every coastal state except Alabama, Mississippi, and Virginia voted for the resolution.

[43] On the legislative history, see Senate, *Estuarine Areas*, pp. 2–4; House, *Estuarine Areas*, pp. 4–9. For Massachusetts' wetland laws, see Foster, "Wetlands Protection Program," pp. 155–63; Reitze, *Environmental Planning*, §2, pp. 60–66.

Absenteeism was a primary factor contributing to the defeat of Dingell's resolution. Nine representatives from California who had expressed support for the measure failed to vote. In addition, four major supporters of either the administration or conservation measures did not vote. These included Claude Pepper of Florida, Wayne Aspinall of Colorado, and Morris Udall of Arizona, all advocates of protection, and Wayne Hayes of Ohio. All told the *Congressional Quarterly* that they would have voted for the measure had they been present. The bill required presidential approval for acquisition of land by the secretary of the interior, but despite this check on the secretary's power, Republicans objected to the program as a usurpation of state and local powers by the federal bureaucracy. Congressman Durward Hall, a Missouri Republican, led the opposition.[44]

Despite the defeat, Dingell and his faction introduced a number of bills on January 10, 1967, differing from the previous resolution but still authorizing tidelands acquisition by the secretary with approval of the president. House hearings were held in late 1967, and testimony was heard by the Commerce Committee to amend the bill and recommend the House's passage. On February 8, 1968, the House passed a Senate-drafted version requiring congressional approval of any future federal acquisition of wetlands besides the Long Island site.[45]

During the course of the hearings in the House, opposition came from the federal bureaucracy, states' rights advocates who viewed the bill's acquisition provisions as an infringement on state police powers, and Republicans who felt it threatened local control and the rights of property owners. Within the government agencies, the Army Corps of Engineers objected to a dual-permit system for the dredging and filling of estuaries as an infringement of its authority and recommended that the legislation be limited to a study and inventory of the nation's estuaries. The Corps of Engineers also criticized the inclusion of the fresh waters of the Great Lakes in the study, as these were not strictly estuaries. The Army was joined by the Department of Housing and Urban Development in trying to limit the bill to a two-year study. These suggestions would have hindered the passage of the legislation in two other respects. The inclusion of the Great Lakes assured that thirty states would be covered by the legislation instead of only the twenty-four tidewater states. This maneuver encour-

[44] *Congressional Quarterly Almanac* 22 (1966): 654, 926–27.
[45] Senate, *Estuarine Areas*, p. 2; House, *Estuarine Areas*, pp. 1–9.

aged support for the measure in both the populous lakeshore districts represented in the House and by a majority of states in the Senate. Secondly, limiting the resolution to a study would have evoked opposition from the Bureau of the Budget, which suggested that the study was a duplication of the $3 million water-pollution study previously authorized by the Water Pollution Control Act. The secretary of the interior and the bill's supporters in the House strongly disagreed, explaining that the study envisioned a broader approach to estuarine protection, including a system of nationwide management policies beyond mere water-pollution control.[46]

During the late spring and early summer of 1967, Stewart Udall moved to blunt the opposition of the Corps of Engineers, and on July 13 he and the secretary of the Army issued a joint memorandum of understanding. It assured full cooperation between the two departments in assessing the broad effects of dredge-and-fill permits on the pollution and biological viability of estuarine areas in addition to the refuse-control authority over navigable waters—a Corps responsibility since 1899. Policies were established for district engineers to alert all interested agencies within the federal and state bureaucracies whenever applications "for dredging, filling, excavation or other related work in navigable waters" were filed with the Army Corps of Engineers. The agreement's legal authority rested on water pollution control legislation and the fish and wildlife coordination acts as amended. Where disagreements existed between agencies, the final resolution would reside jointly with the secretaries of the Army and the interior.[47]

This agreement produced a new draft of the legislation submitted by the Department of the Interior on August 2, 1967. The new resolution, supported by all parts of the federal bureaucracy, was amended during House debates to limit acquisition powers to the Long Island area. Subsequent purchases had to await the end of a two-year study on the feasibility and cost of a federal coastal-wetlands system. After the House passed the amended measure in February, it went to the upper house. The Senate Commerce Committee, under the chairmanship of Warren Magnuson of Washington, reported the bill out with one amendment on July 17, 1968. The committee recommended cutting the appropriations for the two-year study by half. The new version passed the Senate by a voice vote on the nineteenth and the House concurred with the $500,000 appropriation limi-

---

[46] House, *Estuarine and Wetlands Legislation*, pp. 4–19, 70–73, 89–111.
[47] *Congressional Quarterly Almanac* 22 (1966): 654; Senate, *Estuarine Areas*, pp. 14–16.

tation on the twenty-second. Signed into law by President Lyndon Johnson on August 3, 1968, the National Estuary Protection Act committed the nation to the fulfillment of a formidable and extensive challenge.[48]

The long delay in the bill's enactment left only a year and a half for the Interior Department to survey the nearly ninety thousand miles of national shoreline for the purpose of recommending sites for inclusion in a nationwide system of estuarine sanctuaries. Although Congress retained the power to approve future acquisitions and appropriate additional funds for sharing the cost of estuarine-management programs with the states, the growing nationwide consensus for coastal regulation assured that the findings of the Interior Department would gain additional support. Between the passage of the act and the submittal of the report to Congress in January, 1970, the pace of coastal population growth, pollution, and real-estate subdivisions kept the issues raised by New York, Massachusetts, and San Francisco Bay in the national political arena. Confrontation continued on the judicial and state-legislative levels.[49]

Two sections of the initial study were authorized by the bill. The first portion was to identify areas for conservation and protection, while the more detailed section was to suggest the most efficacious means for preserving, protecting, and restoring the nation's estuaries. When the report was submitted in 1970, it considered a variety of means to further the act's intention of balancing development with conservation in the estuarine zone. Court injunctions for cease-and-desist notices to parties engaged in unauthorized filling, dredging, or polluting were one means suggested. Local zoning for open-space or flood-plain protection was recommended, with recourse to federal acquisition only if state and local controls became inadequate.[50]

Other significant provisions of the National Estuary Protection Act strengthened the power of the secretary of the interior, short of giving him outright the ability to acquire property without legislative approval. He was authorized to propose to Congress nationally significant land and water areas within estuaries as part of a joint federal-state system of es-

[48] Senate, *Estuarine Areas*, pp. 1–2, 8, 16; *Congressional Quarterly Almanac* 24 (1968): 311–12; "National Estuary Protection Act," Public Law 90-454, 16 *United States Code Annotated* 1221; *Congressional Record*, CXIV, 2736, 2750–58, 2632.

[49] "National Estuary Protection Act," 82 *Statutes at Large* 628, §2, pp. 2–1; U.S. Department of the Interior, Fish and Wildlife Service, *National Estuary Study*, 7 vols.

[50] House, *Estuarine Areas*, pp. 40–41; Department of the Interior, *National Estuary Study*, VI, 1–80; "National Estuary Protection Act," §2(b), §2(c), §2(d).

tuarine management. The only area Congress had authorized him to acquire immediately was the Long Island wetlands of the Great South Bay. The secretary could order other federal take-over only of public land already owned by local or state entities. He was empowered to create management agreements that private owners could voluntarily enter into, but they could not be required to do so. Other federal agencies were forbidden to dredge, fill, or excavate in estuaries or to authorize such plans without prior approval from the secretary of the interior. Finally, the secretary was to issue regulations governing dredge spoils, earth, refuse, and garbage disposal in estuaries in order to supplement the pollution-abatement guidelines for interstate waters and the navigational directives of the Army Corps of Engineers. The legal power of the secretary rested on four additional acts, the Federal Water Pollution Control Act, the Fish and Wildlife Coordination Act, the Endangered Species Protection Act, and the Anadromous Fisheries Act. The long tradition of piecemeal responsibility hedged in by strict construction of separate agency powers over coastal land and water resources was apparently over.[51]

Introducing his version of the bill in 1967, Senator Edward Kennedy had explained, "This bill is not intended to set up a system of locked up federal areas." Instead, he and Dingell had sought an institutional mechanism for balancing the private desire to develop the seashore against the public trust inherent in navigational servitude; fishery, wildlife, and water-quality protection; state police powers; and the residual interest of the states in ownership of tidelands. Their action on behalf of local conservation coalitions and national preservationist organizations allowed Dingell, Udall, and Magnuson to focus national attention on the need for regulation, coordination, and equity in decisions affecting the biological and cultural integrity of estuaries. Coastal wetlands, as Kennedy had suggested, "are every bit as much a part of our natural heritage as are mountains and great rivers, every bit as deserving of national attention and treatment."[52] Such sentiments represented an enhanced respect for the dependence of fisheries, recreation, and scientific understanding on the biological integrity of estuarine ecosystems. Since the functional integrity of estuaries and long-term economic stability are interrelated through the growth of the country's housing, commercial, and transportation demands, wetlands management requires an extraordinary degree of cooperation.

[51] "National Estuary Protection Act," §3, §4, §5, pp. 2–3.
[52] Senate, *Estuaries and Their Natural Resources*, pp. 2–4.

The estuary protection act recognized this necessity of trilevel planning among national, regional, and local authorities. Under its provisions, federal and state governments shared the funding and review functions involved in the planning process. Municipalities and regional commissions were encouraged to submit water- and land-use plans for estuarine areas that reflected the economic, social, and environmental realities in their respective areas and the effects of state or federal acquisition of coastal wetlands on comprehensive river-basin plans. The intent of the legislation was clear from Kennedy's remark that "these estuarine areas must be treated as one, for the purposes of conservation." [53]

The Senate had broadly defined the meaning of estuarine areas to include "transition zones—salt meadows, coastal marshes, intertidal areas, sounds and other coastal water areas, plus the vital fresh water areas above the limit of salt water intrusion, so important as spawning and nursery areas for many anadromous fish." The breadth of the act's definition and its inclusion of the Great Lakes in the study meant that virtually all shorelines, tidal and non-tidal, were covered by the legislation. Especially impressive to the Senate Commerce Committee had been the report of the Environmental Pollution Panel of the President's Science Advisory Committee concerning the relationship of tidal marshes to estuarine productivity. In the report the committee said that the "peculiar biological value of the estuarine zone" was due in part to "extraordinarily fertile . . . salt marshes." The report referred to Odum's studies that showed marshes "produce nearly seven times as much organic matter per unit area as the water of the Continental Shelf" and "six times as much as average wheat producing land." While the committee recognized regional differences, it noted the dependence on estuarine resources of the Gulf shrimp fisheries, the Pacific Northwest salmon fisheries, and the Atlantic shell fisheries. This geographic diversity helped garner support from representatives of all coastal states. [54]

The trilevel responsibility for planning and funding, the comprehensive inventory, and the assessment of environmental impacts of federal programs and private developments on estuarine environments all contributed to making the estuary protection act what Tenzer referred to as "a new concept" in resource legislation. For this reason the National Estuary Protection Act can be viewed as the forerunner of the National Environmental

[53] Ibid.
[54] Senate, *Estuarine Areas*, pp. 3–5.

Policy Act of 1969 (NEPA) and the Coastal Zone Management Act of 1972 (CZM).[55] More generally, the estuary bill and the report it mandated kept the issues of public trust and intergovernmental responsibility clearly before the courts, media, and state legislatures at a time when several coastal disasters were spotlighting the dangers in coastal development. This review of developmental impacts on the estaurine zones meant that the days of laissez-faire exploitation of the coast were numbered. The effect of the act on the federal courts was reflected in a 1970 district-court ruling to desist from filling and dredging in Boca Ciega Bay off Saint Petersburg, Florida.[56]

Passage of the act and its subsequent study provoked a plethora of articles, books, and essays reflecting a growing cultural appreciation for what had always been considered "wastelands." In November, 1969, *Audubon Magazine* published an edition dedicated to marshland protection, which included a condensation of John and Mildred Teal's forthcoming book, *Life and Death of the Salt Marsh*. The Teals, who had studied under Odum at the Sapelo Island Marine Institute from 1955 until 1969, had eloquently presented these least understood portions of the littoral. Their work centered on the geologic history of East Coast marshlands and the subsequent developments along them. The Teals felt that extensive public efforts were needed to ensure the realization of the goals expressed in federal regulation and that guard could never be relaxed. They warned that "the battle between the forces of development and conservation need be won only once by developers but must be fought and won ever year for conservation to triumph." They insisted that the geographic dimensions of the problem were larger than those of previously protected national parks, monuments, or seashores. "The ribbon of green marshes along the eastern coast of North America . . . must be preserved almost in its entirety if its preservation is to have any real meaning." [57]

Yet another popularizer of the value of marshlands was film executive

---

[55] Tenzer, quoted in *Congressional Quarterly Almanac* 24 (1968): 311; Reitze, *Environmental Planning*, §2, pp. 1–18, 36–59.

[56] *Zabel* v. *Tabb*, 430 *Federal Reporter* 2d, p. 199; L. Eugene Cronin, "The Role of Man in Estuarine Processes," in *Estuaries*, ed. G. H. Lauff, pp. 667, 675–77, 689. Compare the language of "National Estuary Protection Act" (1968) with the intent and wording of "National Environmental Policy Act" (1969), 42 *U.S. Code Annotated* 4321.

[57] John Teal and Mildred Teal, "Ribbon of Green: The Epic of a Salt Marsh," *Audubon Magazine*, November, 1969, p. 4; John Teal and Mildred Teal, *Life and Death of the Salt Marsh*, pp. vi–vii, 261–62.

and naturalist writer Roger Caras, who explained that what many considered smelly marshlands were actually "wetlands that have been badly polluted." He told his readers that a healthy tidal marsh was "a sea-life nursery" and did not produce "that rotten egg smell." [58]

This change in attitudes was also reflected in the writings of Harold Gilliam, who memorialized the fight to protect San Francisco Bay in his 1969 book, *Between the Devil and the Deep Blue Bay*. A long-time critic of what he termed "one-dimensional planning," Gilliam in his articles in Bay-Area daily newspapers informed the northern California region of the growing national consensus for a coastal policy that emphasized environmental values. He lyrically described a salt marsh as "the place where the sea meets the land serenely . . . merging and blending with it to create a rich profusion of life—one of the most abundant types of environment on earth." [59]

Not all citizens were convinced of the newfound dignity for marshlands; one Massachusetts citizen had written protesting his state's legislation: "As the population expands we must have space . . . it becomes a simple matter of priorities." He asked, "Which is going to be destroyed next—the marsh-mudflat or the adjoining meadow or forest?" In his opinion, "to save wetland is to consume dryland," and after years of treating them as wastelands to "believe that swamps are salubrious" was "positively mind bending." Despite such opposition, the popularization of marshland protection based on tidal-marsh productivity received wide play by the media. By 1972, when *Life Magazine* carried a photo essay on tidal marshes entitled "Cradle of Life," the message of a national preservation ideal had become lodged in the popular press. [60]

For all its studies, mandates, and review procedures, the National Estuary Protection Act was more a statement of legislative intent than a tough, enforceable mandate for a new era in man's relation to land and

[58] Roger Caras, "Our Only World: Ugly Monsters in the Swamps," *San Francisco Sunday Chronicle and Examiner*, April 12, 1970.

[59] Harold Gilliam, "Saving the Rare Marshland Beauty of the North Bay," *San Francisco Sunday Chronicle and Examiner*, November 9, 1969.

[60] "Citizen's Letter of Protest," in *Eco-Solutions*, ed. Woods, pp. 164–67; "Cradle of Life," *Life Magazine*, March 17, 1972, p. 34. See also Stephen W. Hitchcock, "Can We Save Our Salt Marshes?" *National Geographic*, June, 1972, pp. 729–65; Mark Jones, "La Ballona Wetlands: Ecosystem vs. the Developers," *Los Angeles Times*, November 17, 1978; Lynda McCormick, "Endangered: L.A.'s Last Marshland," *Christian Science Monitor*, December 28, 1978; Phillip L. Fradkin, "The Mouse That Snored," *Audubon Magazine*, May, 1979, pp. 86–103; "Preserving Suisun Marsh," *Sacramento Bee*, December 31, 1979 (editorial).

water. As conservationists and the estuarine study of January, 1970, pointed out, the real power at the disposal of the federal government to regulate coastal land use was the power of taxation. Like the subsequent CZM, the estuary act of 1968 failed to deal with the problem of property taxes and capital gains taxes, a problem at the very nexus of land, water, and energy policies.[61]

Local reliance on property taxes for schools, police, and fire protection creates opposition from county supervisors when state and federal governments remove developable land from the tax roles for parks. The 50-percent tax allowance on capital gains from the sale of land in the estuarine zone amounts to a federal subsidy of development at the cost of obliterating a national heritage. Furthermore, in coastal zones, where land values increase faster and level off higher than in upland areas, taxes increase accordingly. High taxes encourage owners to subdivide their holdings, complicating coordination and the securing of federal management agreements. Tax benefits or write-offs for owners of undeveloped marshlands might go a long way toward achieving an equitable distribution of the private costs of estuarine preservation.

The problem of equitably sharing the costs of coastal conservation is especially complicated for estuarine resources. The commercial benefits of possessing a healthy tidal marsh are not reaped by the owner. Fishermen on the Continental Shelf are the major direct beneficiaries of preservation, just as they are the losers from heedless pollution and exploitation. Both producers and consumers, on the other hand, benefit from the development of coastal lands for deep-water tanker facilities, liquified natural gas ports, or containerized cargo handling areas. The commercial cost in terms of losses of wetland acreage is borne by fishermen, while costs in terms of air and water pollution associated with street, sewer, and civic improvements are borne by coastal counties. The purchase of scenic easements by state or federal treasuries would be one method of shifting the burden of taxation and conservation from the owners of tidal marshes and local governments—who are ill-equipped to underwrite protection—to those institutions with greater fiscal flexibility. The inability of the taxing system to compensate coastal counties for the deterioration resulting from commer-

[61]Roland C. Clement, "Marshes, Developers and Taxes—A New Ethic for Our Estuaries," *Audubon Magazine*, November, 1969, pp. 34–35; *National Estuary Study*, VI, Appendix 1: "Effects of Taxes and Policies," pp. 1–8.

cial estuarine uses is an essential part of federalism's failure in estuarine protection.

The acquisition of land for wildlife or recreation is not by itself the panacea for coastal and estuarine problems that it has sometimes appeared. First, there are often local objections; second, the cost is prohibitive; and finally, the overuse of recreational facilities degrades the biological integrity of the scenic natural resources.[62] Frequently visited seaside recreation areas are prime contributors to nonpoint sources of pollution and increased erosion and turbidity. User fees as envisioned in the Land and Water Conservation Fund legislation have been less than successful in raising the monies needed to maintain existing recreational facilities and provide for expansion.

Recreation and access rights to publicly owned coastal boating, fishing, camping, and swimming areas have been granted on either a first-come, first-served or an advanced-reservation basis. That constitutional guarantees of access to state or federal lands include provisions for automobiles and off-road vehicles has been a tacit assumption of all concerned. While we have come nearer to Leopold's ideal of a land ethic, the significant change in personal values that it ultimately will require has just begun.[63]

The problem of estuarine management and protection is compounded by the fact that most of the immediate benefits are privately derived from commonly owned resources. The ultimate consequences of the public-goods argument for conservation were addressed by ecologist Garrett Hardin, several months after the passage of the National Estuary Protection Act, in a seminal article entitled "The Tragedy of the Commons." He argued that socially shared resources like air, water, ocean fisheries, or national parks are comparable to publicly owned grazing lands because the costs of maintaining the equitable distribution of public resources escape any technically effective solution. For example, the price for using estuarine resources is shared by all taxpayers, while the returns accrue to those individuals who most often use the estuaries.

Taking Leopold's worries over the costs of conservation to their logical conclusion, Hardin warned that when people seek to maximize their

[62] Roger Revelle, "Outdoor Recreation in a Hyper-Productive Society," *Daedalus* 96 (Fall, 1967): 1172–91.

[63] Simon, *Thin Edge*, pp. 17–19, 83–87, 89–90, 93–96.

own returns there is no immediate, individual incentive to enhance the common good, while mounting cost to the entire group ultimately "brings ruin to all." Ecologically speaking, several commonly owned goods comprise estuarine resources, and their continued use accelerates the degradation of coastal biotic communities. As long as there was an abundance of resources or the population remained small, the productive ecosystems of the estuarine zones were not overburdened. That, however, is no longer the case.[64]

Hardin's remedy for such situations centered on mutually agreed-upon coercion to restrain any one individual from harming the community by indiscriminately furthering his own ends. The basis for governmental action on behalf of the biological and cultural integrity of estuaries is multifaceted. Authority rests on international treaties to protect birds and marine mammals, on constitutional grounds of protecting interstate commerce and promoting the general welfare, and on state police powers to protect the health, safety, and morals of the community. The avenues for coercion are many. Taken together, they have been used to mandate the first nationwide system of land-use planning by the CZM.[65]

Coercing compliance with environmental protection laws through fines and incarceration has proven difficult, litigious, and expensive. Currently, pressure is mounting to shear the Army Corps of Engineers of authority over certain tidal wetlands. This momentous reversal of legislative and judicial intent would seriously disrupt estuarine ecology. By narrowly interpreting §404 of the Clean Water Act, recent federal rules would allow filling and dumping in all areas above the mean high tide at the discretion of private owners. Removal of the Corps's permit-granting powers over this upper marsh zone will increase sedimentation and erosion, alter navigation channels, and remove vital flood-control lands from public scrutiny. The relatively high status accorded tidal wetlands among natural areas by ecologists rests on the indivisibility of each ecosystem's constituent elements for maintaining the physical and biochemical conditions for energy transfer, nutrient recycling, and reproduction. The upper marshes above the mean high tides are as essential to estuarine integrity as clean water. If the §404 review procedure is so prohibitively expensive to waterfront owners that marsh protection is undermined, then the reasonable remedy re-

[64] Garrett Hardin, "The Tragedy of the Commons," *Science*, December 13, 1968, pp. 1243–48.
[65] Ibid., p. 1247.

quires tax revision and not the reaffirmation of laissez-faire reclamation within the estuarine commons.[66]

The *National Estuary Study* (1970) emphasized another dilemma faced in estuarine protection, restoration, and development since the goal is to preserve functions and not merely sequester habitats. Saving the remaining tidelands of the coast will be of little avail if deterioration of riverine water quality destroys the biological integrity of marshes and estuaries. Successful experiments have been conducted involving the replanting of marshes, mudflats, and dredge spoils. In some areas, for example Seattle and San Diego, efforts have been coupled with improvement of estuarine water quality through upstream municipal pollution control.[67]

Today a nationwide system of estuarine sanctuaries exists as a testament to the work of dedicated scientists, preservationists, citizens, and administrators who value the biological equilibrium of scarce open spaces for the sustenance of fish, wildlife, water quality, and waste recycling. These estuarine sanctuaries along the coasts of Oregon, California, Connecticut, Georgia, Texas, and Hawaii are the products of the effective and equitable planning undertaken by local, state, and federal agencies under the estuarine protection act. Amid contending commercial, recreational, and residential shoreline uses, the protection afforded estuarine sanctuaries represents institutional maintenance of necessary physical processes easily destroyed by technological development. Yet public responsibility for equitable estuarine-resource use remains too narrowly interpreted and inadequately funded for coastal conservation to serve fully the pressing demands of ecosystems and citizens alike.

[66] *Federal Register* 47 (January 13, 1982): 1697–98; "Federal Water Pollution Control Act," 33 *United States Code Annotated* 1251; "Lake Ophelia: What Defines a Wetland?" *National Wetlands Newsletter* (March–April, 1981): 15–17.

[67] *National Estuary Study*, IV, Appendix C, pp. 9–11; F. Graham, Jr., "Man Can Create a Marsh," *Audubon Magazine*, September, 1977, p. 140; Lawrence E. Jerome, "Marsh Restoration," *Oceans* 12 (January–February, 1971): 57–59; *New York Times*, September 18, 25, 1970.

9

Americans and the Tidal Seas

> I have seen the hungry ocean gain
> Advantage on the kingdom of the shore,
> And the firm soil win of the watery main,
> Increasing store with loss and loss with store.
> —William Shakespeare, "Sonnet LXIV"

As the American nation has grown older, its attitudes toward its natural resources have changed and matured, and its legal underpinnings for environmental policies have been reinterpreted and reformulated to accommodate the changing understandings and values. The estuarine preservation ideal that emerged in the mid- to late-1960s is one of the most recent developments in American ideas concerning resource stewardship.

Although the environmental protection ideal can claim a heritage from the national concern for fish and wildlife conservation, it is much broader than that. The popularization during the 1950s and 1960s of scientific findings on the role of estuaries in the larger biological scheme of things helped inspire a national constituency to advocate new policies. The new ideal recognizes that healthy tidal wetlands are essential to fisheries, flood control, coastal water supplies, limited sewage processing, and recreation. As freshwater recharge areas, coastal wetlands are necessary for the percolation of fresh surface water into underground aquifers. As the sites of enormous decomposition and recycling, estuaries are integral parts of the earth's ecology. Tidal marshes are supremely suited for flood control and as wildlife habitats. Even as cultural sites, these zones demand special preservation efforts. The historic shores of this nation's bays, sounds, river mouths, and lagoons—for instance, Defauskee and Galveston islands—are layered with the remains of prolonged settlement and graced with surviving historic structures, despite the clutter of the contemporary. The new understanding of wetlands, though, goes beyond such utilitarian motives to a biotic or ecological view of mankind's place in nature. It replaces a notion of land as simply real estate with a concept of land as the living matrix in which energy is converted into life. It acknowledges the ultimate human dependence on healthy biotic relations.

The physical conversion and storage of energy necessary for the production of nucleic acids is at the very center of life's coevolution. Humans share these biochemical genetic processes with all other creatures, from amoebas to sandhill cranes. Short-circuiting the energy available to each creature from natural estuarine cycles for their storage of vital information about physical change through nucleic-acid synthesis and replication diminishes the entire biotic community's continuing adaptive capacity. Ultimately this would deprive coastal life of the physical diversity and coevolution that depends on this integrity of genetic inheritance. Thus, the preservation of estuarine wilderness has come to be considered investment in a phylogenetic savings account.

Moreover, the estuarine preservation ideal is not a closed, dogmatic creed, but rather a mature view of humans as part of an unfinished process. It recognizes that for every action taken to protect some estuary or estuarine creature other losses must occur. The inevitability of such trade-offs was expressed by Shakespeare in "Sonnet LXIV" in his reference to the contest between the "firm soil" and the "watery main," where increase is a reciprocal process of "store with loss *and* loss with store" (emphasis added). The estuarine preservation ideal recognizes that the worth of protecting estuarine integrity must be measured against what the society can afford to do without and what must be protected for the future.

Nonetheless, critics of the National Estuary Protection Act and other policies based on this ideal have persuasively argued that these programs have placed wilderness values ahead of more important human needs for adequate housing, jobs, and raw materials. Local objections to the preservation of coastal wetlands have focused on the practical implications of an estuarine preservation ideal, such as the loss of tax revenues. Since national land-use planning in coastal zones was begun in 1972, private interests have attacked coastal-wetlands protection as the subjection of citizens to interfering, insensitive, centralized authority for the benefit of a few nature enthusiasts.

Until the 1960s important economic and libertarian considerations such as these repeatedly took priority over wildlife preservation; they will continue to challenge the new consensus. When the political coalition that formed in the mid-sixties to assert the intangible as well as long-term economic values of the marshes disintegrates, the more immediate concerns of energy development and housing are likely to overwhelm the enforcement and funding of the estuarine preservation ideal. Nonetheless, the scientific underpinnings, biological significance, and aesthetic appeal of this ideal may make it robust enough to find a place in coastal planning programs.

There are two other threats to continuing preservation of coastal wetlands, however. First, like that of other wild environments, the protection of wetlands cannot be assured by the mere preservation of landscapes, but depends on the sustained quality of the resources that feed into them, like water, and the functional integrity of bordering environments. Second, there is a danger in complacency concerning the responsibility for estuarine preservation. Because tidelands belong to everyone, the responsibility for their maintenance and equitable use remains with public agencies. But the responsibility for protecting water quality and shore access is too diffuse, even when guarantees exist explicitly in state constitutions or implicitly in common law. Presidents and governors are too removed from the problem; Congress and state legislatures are often too clumsy to formulate effective policies; administrative agencies are too fragmented and short-sighted. Responsibility is made more difficult to assess because of the complexity of intergovernmental relations in a federal system.

But responsibility must be fixed, and estuarine preservation must be guaranteed. The shore, as Rachel Carson once reminded America, is "our last outpost," and "the wild seacoast is vanishing." The remaining 260,000 square miles of tidal marshland persist as an important reminder of the collective evolutionary and historical past and as the crib of the future.[1]

"Is civilization progress?" Charles Lindbergh once asked. "The final answer will be given not by our amassment of knowledge or by the discoveries of our science, or by the speed of our aircraft, but by the effects our civilized activities as a whole have upon the qualities of our planet's life— the life of plants and animals as well as that of men."[2] Given the nation's long and crucial reliance upon estuarine resources, the current commitment to stewardship through the retention of wildlife refuges among the

[1] Paul Brooks, *House of Life: Rachel Carson at Work*, p. 225; U.S. Department of the Interior, Fish and Wildlife Service, *National Estuary Study*, II, 122.

[2] Charles Lindbergh, quoted in *Audubon*, July, 1969, p. 34.

tidal marshes represents a societal acceptance of ethical accountability for the natural features and the functioning ecological integrity of the coast.

Restoring biotically healthy estuaries offers human institutions a choice cutting to the heart of democratic values and conservation ideals. We can destroy the balance of extractable resources in estuaries, thereby creating an anaerobic soup out of prolific coastal marshes. But these landscapes will be formed anew with the next geological advance or retreat of the ocean. We can provide sanctuaries for contemplation and repose, or recreation for power boats and dune buggies—or remember to protect prolific fisheries. Ultimately the natural equilibrium of ecosystems forged over millenia of nucleic-acid exchanges will resiliently prevail over human ignorance. In perishing, we can either leave behind a richer more diverse land than we were given or an impoverished landscape devoid of the readily identifiable natural features on which our psychic and economic advance depends.

Reserving significant lands for special social functions in perpetuity is the essence of the ancient public trust. To deny that we owe untold future generations the privilege of watching a great blue heron feed or the right to experience a seal herd basking in the sun is the greatest of follies. It amounts to a denial of an obligation. If nothing else, each generation has a responsibility to its countless predecessors to pass on a qualititatively more valuable legacy. Human dignity requires that the enhancement of natural public areas be every individual's passionate duty. The art of caretaking requires the political courage of citizens to assert the rights of future generations to benefit from the innate fertility of river mouths and to enjoy continued and meaningful contact with the ebb and flood of the sea. Estuaries, a womb of life, are a ready reminder of the ethical distance we have come and have yet to travel.

In these coastal regions where American civilization has tread most heavily, wild marshes hold special appeal. The untamed beauty and wide expanse of those quiet shores sing to modern culture a haunting refrain in the never-ending yet urgent search by humans to find their identity in relation to natural landscapes. These few tidal marshes remain to admonish us that we are the current caretakers of a delicately coevolved maritime garden whose produce must always be thoughtfully harvested.

Selected Bibliography

Books and Articles

Agassiz, Elizabeth C., and Alexander A. Aggassiz. *Seaside Studies in Natural History*. Boston: Ticknor and Fields, 1865.

Agassiz, G. R., ed. *Letters and Recollections of Alexander A. Aggassiz, with a Sketch of His Life and Work*. Boston: Houghton Mifflin, 1913.

Allard, Dean Conrad, Jr. *Spencer Fullerton Baird and the United States Fish Commission*. New York: Arno Press, 1978.

Allee, Warder C.; Alfred E. Emerson; Orlando Park; Thomas Park; and Karl P. Schmidt. *Principles of Animal Ecology*. Philadelphia: W. B. Saunders, 1949.

Allen, J. "Hydraulic Engineering." In *History of Technology*, edited by Joseph Singer. 10 volumes. Oxford: Clarendon Press, 1958.

Ayres, Quincy, and Daniel Scoates. *Land Drainage and Reclamation*. New York: McGraw-Hill, 1928.

Baily, Francis. *Journal of a Tour in Unsettled Parts of North America in 1796 and 1797*. Edited by Jack D. L. Holmes. Carbondale: University of Southern Illinois Press, 1969.

Bakeless, John. *The Eyes of Discovery*. New York: Dover Publications, 1961.

Banta, John S. "Constitutional Issues and Estuarine Management." *Oceanus* 19 (Fall, 1976): 64–70.

Barnes, R. S. K., and J. Green, eds. *The Estuarine Environment*. London: Applied Science Publishers, 1972.

Bartley, Ernest R. *The Tidelands Oil Controversy*. Austin: University of Texas Press, 1953.

Bartram, William. *The Travels of William Bartram*. Edited by Francis Harper. New Haven: Yale University Press, 1958.

Bascom, Willard. "Beaches." *Scientific American*, August, 1960, pp. 1–12.

Baumhoff, Martin. *Ecological Determinants of Aboriginal California Populations*. University of California Publications in American Archaeology and Ethnology, Vol. 49, No. 2. Berkeley: University of California Press, 1963.

Beard, Charles A., ed. *A Century of Progress*. New York: Harper and Row, 1932.

Becker, Carl. *Beginnings of the American People*. Ithaca: Cornell University Press, 1915.

Beverly, Robert. *The History of the Present State of Virginia*. Edited by Louis B. Wright. 1705; reprint ed., Chapel Hill: University of North Carolina Press, 1947.

Bowden, Charles. *Killing the Hidden Waters*. Austin: University of Texas Press, 1977.

Brewer, Richard. "A Brief History of Ecology: Pre-nineteenth Century to 1910." *Occasional Papers of the C. C. Adams Center for Ecological Studies*, No. 1 (November 22, 1960). Kalamazoo: Western Michigan University.

Bridenbaugh, Carl. *Cities in the Wilderness: The First Century of Urban Life In America, 1625–1742*. New York: Ronald Press, 1938; reprint ed., New York: Alfred Knopf, 1955.

Brooks, Paul. *The House of Life: Rachel Carson at Work*. Boston: Houghton Mifflin, 1972.

Brown, Ralph H. *Historical Geography of the United States*. New York: Harcourt Brace, 1948.

————. "The Land and the Sea: Their Larger Traits," *Annals of the Association of American Geographers* 41 (September, 1951): 199–209.

Bruce, Phillip A. *Economic History of Virginia in the Seventeenth Century*. New York: Macmillan, 1896.

Bryant, William Cullen. *Picturesque America: Or The Land We Live In*. 2 vols. New York: Appleton, 1872.

Burton, Maurice. *Margins of the Sea*. New York: Harper Brothers, 1954.

Callison, Charles H., ed. *America's Natural Resources*. New York: Ronald Press, 1957.

Carr, Donald E. *Death of the Sweet Waters*. New York: Norton, 1966.

Carson, Rachel. *The Edge of the Sea*. New York: New American Library, 1955.

————. "Our Ever Changing Shore." *Holiday Magazine*, July, 1958.

————. *The Sea Around Us*. New York: Oxford University Press, 1950.

————. *Silent Spring*. Boston: Houghton Mifflin, 1962.

————. *Under the Seawind*. New York: Simon and Schuster, 1941.

Clarke, Robert. *Ellen Swallow: The Woman Who Founded Ecology*. Chicago: Follett Publishing, 1973.

Clepper, Henry, ed. *Origins of American Conservation*. New York: Ronald Press, 1966.

Cloverdale, Joan. *I Share This Marsh: A Poetic Reflection*. Newport Beach, Calif.: Friends of Newport Bay, 1976.

Coker, R. E. *This Great and Wide Sea: An Introduction to Oceanography and Marine Biology*. New York: Harper and Row, 1947.

Commoner, Barry. *The Closing Circle: Nature, Man and Technology*. New York: Alfred Knopf, 1971.

Connery, Robert H. *Governmental Problems in Wildlife Conservation*. New York: Columbia University Press, 1935.

Conron, John. *The American Landscape*. New York: Oxford University Press, 1974.

Costa, Daniel. "The Sea Otter: Its Interaction with Man." *Oceanus* 21 (Spring, 1978): 24–30.

Craven, Avery Odelle. *Soil Exhaustion as a Factor in the Agricultural History of Virginia and Maryland, 1606–1860*. Urbana: University of Illinois Press, 1925.

Cronin, L. Eugene. "The Role of Man in Estuarine Processes." In *Estuaries*, edited by G. H. Lauff. A.A.A.S. Publication No. 83. Washington, D.C.: American Academy for the Advancement of Science, 1967.

Currie, William. "An Inquiry into the Causes of the Insalubrity of Flat and Marshy Situations; and Directions for Preventing or Correcting the Effects Thereof." *Transactions of the American Philosophical Society* 4 (1799): 127–42.

Curti, Merle. *American Paradox: The Conflict of Thought and Action*. New Brunswick: Rutgers University Press, 1956.

————. *The Growth of American Thought*. New York: Harper and Row, 1964.

Dall, William Healey. *Spencer Fullerton Baird: A Biography*. Philadelphia: J. B. Lippincott, 1915.

Daniels, George H. *American Science in the Age of Jackson*. New York: Columbia University Press, 1968.

————, ed. *Nineteenth Century American Science: A Reappraisal*. Evanston: Northwestern University Press, 1972.

Darrah, William Culp. *Powell of the Colorado*. Princeton: Princeton University Press, 1951.

Davis, John H. "Influences of Man upon Coast Lines." In *Man's Role in Changing the Face of the Earth*, edited by William L. Thomas, Jr. Chicago: University of Chicago Press, 1956.

Deacon, Margaret. *Scientists and the Sea, 1650–1900: A Study of Marine Science*. London: Academic Press, 1971.

Doughty, Robin W. *Plume Birds and Feather Fashions: A Study of Nature Protection*. Berkeley and Los Angeles: University of California Press, 1975.

Driver, Harold. *Indians of North America*. Chicago: University of Chicago Press, 1969.

Dupree, A. Hunter. *Science in the Federal Government*. Cambridge, Mass.: Belknap Press, 1957.

Erikson, Erik. *Family, Childhood, and Society*. New York: W. W. Norton, 1963.

Fanning, J. T. *A Practical Treatise on Hydraulic and Water Supply Engineering*. New York: Van Nostrand, 1902.

Farrington, Benjamin. *Francis Bacon: Philosopher of Industrial Science*. New York: H. Schuman, 1949.

Flader, Susan. *Thinking Like a Mountain: Aldo Leopold and the Evolution of an Ecological Attitude toward Deer, Wolves, and Forests*. Missouri: University of Missouri Press, 1974.

Fleming, Donald. "The Roots of the New Conservation Movement." *Perspectives in American History* 6 (1972): 7–91.

Gabrielson, Ira. *Wildlife Refuges*. New York: Macmillan, 1943.

Garret, Edmund H. *Romance and Reality of the Puritan Coast*. Boston: Little Brown, 1897.

Gates, David Allen. *Seasons of the Salt Marsh*. Old Greenwich, Conn.: Chatham Press, 1975.

George, Henry. "Our Land and Land Policy" (1871). *The Writings of Henry George*. Vol. 9. New York: Doubleday and McClure, 1901.

————. *Progress and Poverty*. New York: D. Appleton, 1881.

Gilbert, Arlan K., ed. "Oliver Evans' Memoir 'On the Origin of Steam Boats and Steam Waggons.'" *Delaware History* 7 (September, 1956): 142–67.

Gilliam, Harold. "The Fallacy of Single-Purpose Planning," *Daedalus* 96 (Fall, 1967): 1142–57.

————. "Saving the Rare Marshland Beauty of the North Bay." *San Francisco Sunday Examiner and Chronicle*, November 9, 1969.

Glacken, Clarence. "The Origins of the Conservation Philosophy." *Journal of Soil and Water Conservation* 2 (1956): 53–56.

————. *Traces on the Rhodian Shore*. Berkeley and Los Angeles: University of California Press, 1972.

Goldberg, Edward. "Pollution History of Estuarine Sediments." *Oceanus* 19 (Fall, 1976): 18–26.

Graham, Otis L., Jr. *Toward a Planned Society*. New York: Oxford University Press, 1976.

Grodzins, Morton, and Eugene Rabinowitch, eds. *The Atomic Age: Scientists in National and World Affairs*. New York: Simon and Schuster, 1963.

Gunter, Gordon. "Some Relations of Faunal Distributions to Salinity in Estuarine Waters." *Ecology* 37 (July, 1956): 616–19.

Haeckel, Ernst. *Generelle Morpologie der Organismen*. Berlin, 1866.

Hanie, Robert. *Guale, The Golden Coast of Georgia*. New York: Seabury Press, 1974.

Hakluyt, Richard. *Hakluyt's Voyages: The Principal Navigations, Voyages, Traffiques and Discoveries of the English Nation*. Edited by Irwin R. Blacker. London: George Bishop, 1600; reprint ed., New York: Viking Press, 1965.

Hardin, Garrett. "The Tragedy of the Commons." *Science*, December 13, 1968, pp. 1243–48.

————, and John Baden, eds. *Managing the Commons*. San Francisco: W. H. Freeman, 1977.

Hardy, Sir Allister. *The Open Sea: Its Natural History*. Reprint, 2 vols. in 1. Boston: Houghton Mifflin, 1965.

Harris, L. E. *Vermuyden and the Fens: A Study of Sir Cornelius Vermuyden and the Great Level*. London: Cleaver-Hume Press, 1953.

Hays, Samuel P. *Conservation and the Gospel of Efficiency: The Progressive Conservation Movement, 1890–1920*. Cambridge: Harvard University Press, 1959.

Healy, Robert G. *Land Use and the States*. Baltimore: Johns Hopkins University Press, 1979.

Hedgpeth, Joel. *Treatise on Marine Ecology and Paleoecology*. Washington, D.C.: Geological Society of America, 1957.

————. "Voyage of the Challenger." *Scientific Monthly* 63 (September, 1946): 194–202.

————, ed. *Between Pacific Tides*. Rev. ed. Stanford: Stanford University Press, 1968.

Herdman, William A. *The Founders of Oceanography and Their Work: An Introduction to the Science of the Sea*. London: Edward Arnold, 1923.

————. "Oceanography, Bionomics and Aquiculture." *Report of the British Association for the Advancement of Science* 52 (September 19, 1895). Reprinted in *Man and the Sea: Classic Accounts of Marine Exploration*, edited by Bernard L. Gordon. Garden City, N.Y.: Doubleday, 1972.

Hewlett, Richard G., and Francis Duncan. *Nuclear Navy, 1946–1962*. Chicago: University of Chicago Press, 1974.

Heyman, Ira Michael. "The Great 'Property Rights' Fallacy." *Cry California* 3 (Fall, 1968): 29–34.

Hibbard, Benjamin Horace. *A History of Public Land Policies*. New York: Macmillan, 1924.

Hindle, Brooke. *The Pursuit of Science in Revolutionary America*. New York: W. W. Norton, 1956.

Hitchcock, Stephen W. "Can We Save Our Salt Marshes?" *National Geographic*, June, 1972.

Hofstadter, Richard. *The American Political Tradition*. New York: Alfred Knopf, 1948.

Hollister, Warren. *Medieval Europe*. New York: John Wiley and Sons, 1968.

Hosmer, Charles H. *The Presence of the Past: A History of the Preservation Movement*. New York: G. P. Putnam, 1965.

Howe, Henry F. *Salt Rivers of the Massachusetts Shore*. New York: Rinehart, 1951.

Huth, Hans J. *Nature and the American: Three Centuries of Changing Attitudes*. Berkeley and Los Angeles: University of California Press, 1957.

Ingle, Robert M. "The Life of an Estuary." *Scientific American*, May, 1956, pp. 64–68.

Institute of Ecology. *Man in the Living Environment: A Report on Global Ecological Problems*. Madison: University of Wisconsin Press, 1972.

Ise, John. *United States Forest Policy*. New Haven: Yale University Press, 1920.

Jenks, Cameron. *The Bureau of the Biological Survey*. Baltimore: Johns Hopkins University Press, 1929.

Kammen, Michael. *People of Paradox: An Inquiry Concerning the Origins of American Civilization*. New York: Alfred Knopf, 1972.

Kastner, Joseph. *A Species of Eternity*. New York: E. P. Dutton, 1977.

Kelly, Alfred H., and Winfred A. Harbison. *The American Constitution: Its Origins and Development*. New York: W. W. Norton, 1963.

Kemble, John Haskell. *San Francisco Bay: A Pictorial Maritime History*. New York: Bonanza Books, 1947.

Kidson, C. "Coastal Conservation in Great Britain." *Geography* 229 (July, 1964): 314–22.

Kiernan, John. *A Natural History of New York City*. Boston: Houghton Mifflin, 1959.

Kirker, Harold. *California's Architectural Frontier*. Santa Barbara: Peregrine Smith, 1973.

Kniffen, Fred B. *Pomo Geography*. University of California Publications in American Archaeology and Ethnology, Vol. 36 (1935–1939). Berkeley: University of California Press, 1940.

Kohlstedt, Sally Gregory. *The Formation of the American Scientific Community*. Chicago: University of Illinois Press, 1976.

Lambert, Audrey M. *The Making of the Dutch Landscape: An Historical Geography of the Netherlands*. London: Seminar Press, 1971.

Landsburg, Hans H. *Natural Resources for U.S. Growth*. Baltimore: Johns Hopkins University Press, 1964.

Lanier, Sidney. "The Marshes of Glynn." In *American Poetry and Prose: Part II, Since the Civil War*, edited by Norman Foerster. Boston: Houghton Mifflin, 1934.

Lankford, John, ed. *Captain John Smith's America*. New York: Harper and Row, 1967.

Lauff, George, ed. *Estuaries*. A.A.A.S. Publication No. 83. Washington, D.C.: American Academy for the Advancement of Science, 1967.

Le Duc, Thomas. "The Historiography of Conservation." *Forest History* 9 (October, 1965): 23–28.

Leopold, Aldo. *A Sand County Almanac: With Essays on Conservation from Round River*. New York: Sierra Club/Ballantine, 1966.

———. *Round River: From the Journals of Aldo Leopold*. Edited by Luna B. Leopold. New York: Oxford University Press, 1953.

Lindeman, Raymond. "The Trophic Dynamic Aspect of Ecology." *Ecology* 23 (1942): 399–418.

Lowenthal, David. *George Perkins Marsh: Versatile Vermonter*. New York: Columbia University Press, 1958.

Lunt, Dudley Cammett. *Thousand Acre Marsh: A Span of Remembrance*. New York: Macmillan, 1959.

Lurie, Edward. *Louis Agassiz: A Life in Science*. Chicago: University of Chicago Press, 1960.

———. *Nature and the American Mind*. New York: Science History Publications, 1974.

Lyell, Sir Charles. *Principles of Geology*. 3 vols. London: John Murry, 1830–33.

Maass, Arthur. *Muddy Waters: The Army Corps of Engineers and the Nation's Rivers*. Cambridge: Harvard University Press, 1951.

Malthus, Thomas. "A Summary View of the Principle of Population." In *Introduction to Malthus*, edited by D. V. Glass. London: Watts, 1953.

Marsh, Caroline Crane. *The Life and Letters of George Perkins Marsh*. Vol. 1. New York: Charles Scribner's, 1888.

Marsh, George Perkins. *Man and Nature: Or Physical Geography as Modified by Human Action*. Edited by David Lowenthal. Cambridge, Mass.: Belknap Press, 1965.

Marx, Wesley. *The Frail Ocean*. New York: Ballantine, 1967.

Maury, Matthew Fontaine. *The Physical Geography of the Sea*. New York: Harper Brothers, 1855.

McLanathan, Richard. *Art in America: A Brief History*. New York: Harcourt Brace Jovanovich, 1973.

Merriman, Mansfield. *Treatise on Hydraulics*. New York: Wiley and Sons, 1907.

Miller, G. Tyler, Jr. *Living in the Environment: Concepts, Problems, and Alternatives*. Belmont, Calif.: Wadsworth Publishing, 1975.

Miller, Perry. *Nature's Nation*. Cambridge, Mass.: Belknap, 1967.

Morison, Samuel Eliot. *The European Discovery of America: The Northern Voyages*. New York: Oxford University Press, 1971.

———. *A Maritime History of Massachusetts*. Boston: Houghton Mifflin, 1921.

Muir, John, ed. *Picturesque California and the Region West of the Rocky Mountains from Alaska to Mexico*. 2 vols. San Francisco: J. Dewing, 1888.

Mumford, Lewis. *The Brown Decades: A Study of the Arts in America, 1865– 1895*. New York: Dover Publications, 1955.

———. *The City in History: Its Origins, Its Transformations, and Its Prospects*. New York: Harcourt, Brace and World, 1961.

———. *Sticks and Stones: A Study of American Architecture and Civilization*. New York: Dover Publications, 1955.

———. *Technics and Civilization*. New York: Harcourt, Brace and World, 1934.

Nash, Roderick. "The American Conservation Movement." In *Forums in History*. St. Charles, Mo.: Forum Press, 1974.

———. *Wilderness and the American Mind*. New Haven: Yale University Press, 1967.

———, ed. *The American Environment*. Reading, Mass.: Addison Wesley, 1976.

Newell, Frederick. "The Engineers' Part in After-the-War Problems," *Scientific Monthly* 8 (March, 1919): 239–46.

———. "What May Be Accomplished by Reclamation." *Annals of the American Academy of Political and Social Science* 33 (1909): 658–63.

Niering, William A. "The Dilemma of Coastal Wetlands: Conflict in Local, National and World Priorities." In *The Environmental Crisis*, edited by Harold W. Helfrich, Jr. New Haven: Yale University Press, 1970.

Odgers, Merle. *Alexander Dallas Bache: Scientist and Educator*. Philadelphia: University of Pennsylvania Press, 1947.

Odum, Eugene P. *Fundamentals of Ecology*. Philadelphia: W. B. Saunders, 1971.

———. "The New Ecology." *Bioscience* 16 (July, 1964): 14–16.

———. "The Pricing System," *Georgia Conservancy Magazine* (4th quarter, 1973): 8–10.

———. "The Role of Tidal Marshes in Estuarine Production," *Conservationist* (June–July, 1961): 12–35.

Odum, Howard T. *Environment, Power, and Society*. New York: Wiley-Interscience, 1971.

Olmsted, Frederick Law, Sr. *The Papers of Frederick Law Olmsted*. Volume 1. Edited by Charles Capen McLaughlin. Baltimore: Johns Hopkins University Press, 1978.

———. "Public Parks and the Enlargement of Towns." *Journal of Social Science* 3 (1871): 1–36.

Oosting, Henry J. *The Study of Plant Communities*. San Francisco: Freeman, 1948.

Parr, Charles McKew. *The Voyages of David de Vries: Navigator and Adventurer*. New York: Thomas Y. Crowell, 1969.

Parrington, Vernon Louis. *Main Currents in American Thought*. Vol. 2, *The Romantic Revolution*. New York: Harcourt, Brace and World, 1927.

Passmore, John. *Man's Responsibility for Nature: Ecological Problems and Western Traditions*. New York: Charles Scribner's Sons, 1974.

Pearse, A. S. *The Emigrations of Animals from the Sea*. New York: Sherwood Press, 1950.

Petulla, Joseph M. *American Environmental History*. San Francisco: Boyd and Fraser, 1977.

Phillips, Ulrich Bonnell. *Life and Labor in the Old South*. Boston: Little Brown, 1924.

Pinchot, Gifford. *The Fight for Conservation*. Seattle: University of Washington Press, 1967.

Pirenne, Henri. *Medieval Cities*. Princeton: Princeton University Press, 1925.

Pomeroy, Earl. *The Pacific Slope*. Seattle: University of Washington Press, 1965.

Postel, Mitchell. "The Legacy of a Lost Resource: The History of the Fishing Industry off the San Mateo County Bay Line." *San Mateo County Historical Museum Papers* (May, 1978): 1–57.

Powell, John Wesley. "Institutions for the Arid Lands." *Century Illustrated Monthly Magazine* 40 (1890): 111.

Pursell, Carroll W., Jr. *Readings in Technology and American Life*. New York: Oxford University Press, 1969.

Rakestraw, Lawrence. "Conservation Historiography." *Pacific Historical Review* 42 (1972): 271–88.

Ranwell, D. S. *Ecology of Salt Marshes and Sand Dunes*. London: Chapman Hall, 1972.

Reitze, Arnold, ed. *Environmental Planning: Law of Land and Resources*. Washington, D.C.: North American International, 1974.

Revelle, Roger. "Outdoor Recreation in a Hyper-Productive Society." *Daedalus* 96 (Fall, 1967): 1172–91.

Ripley, S. Dillon, and Helmut K. Buechner, "Ecosystem Science as a Point of Synthesis." *Daedalus* 96 (Fall, 1967): 1192–99.

Robbins, Roy Marvin. *Our Landed Heritage: The Public Domain, 1776–1936*. Princeton: Princeton University Press, 1942.

Rockefeller Commission Report. *Citizen's Guide to Land-Use Planning*. Edited by W. K. Reilly. New York: Thomas Y. Crowell, 1973.

Rohrbough, Malcolm J. *Land Office Business: The Settlement and Administration of American Public Lands, 1789–1837*. New York: Oxford University Press, 1968.

Roper, Laura Wood. *Frederick Law Olmsted: A Biography*. Baltimore: Johns Hopkins University Press, 1974.

Rose, Edward J. *Henry George*. New York: Twane Publishing, 1968.

Ruffin, Edmund. *Agricultural, Geological, and Descriptive Sketches of Lower North Carolina*. Raleigh, N.C.: Institution for the Deaf and Dumb and the Blind, 1861.

———. *Essays and Notes on Agriculture*. Richmond: J. W. Randolf, 1855.

Rush, Benjamin. "An Inquiry into the Cause of the Increase of Billious and Inter-mitting Feveers, in Pennsylvania, with Hints for Preventing Them." *Transactions of the American Philosophical Society* 2 (1786): 205–209.

Sauer, Carl Ortwin. *Land and Life*. Edited by John Leighly. Berkeley and Los Angeles: University of California Press, 1963.

———. *Sixteenth Century North America*. Berkeley and Los Angeles: University of California Press, 1971.

Scharf, J. Thomas. *The Chronicles of Baltimore: Being a Complete History of "Baltimore Town" and Baltimore City from the Earliest Period to the Present Time*. Port Washington, N.Y.: Kennickat Press, 1972.

Schlee, Susan. *The Edge of an Unfamiliar World*. New York: E. P. Dutton, 1973.

Scott, Mel. *American City Planning*. Berkeley and Los Angeles: University of California Press, 1971.

———. *The Future of San Francisco Bay*. Berkeley: Institute of Governmental Affairs, University of California, 1963.

Scott, Stanley. "Coastal Planning in California: A Progress Report." *Bulletin of the Institute for Governmental Affairs*, vol. 19. Berkeley: University of California, 1963.

Sears, Paul B. *Where There Is Life: An Introduction to Ecology*. New York: Dell Publishers, 1962.

Seybert, Adam. "Experiments and Observations on the Atmosphere of Marshes." *Transactions of the American Philosophical Society* 4 (1799): 415–28.

Shaler, Nathaniel Southgate. *Man and the Earth*. New York: Fox Duffield, 1905.

———. *Sea and Land*. New York: Charles Scribner's Sons, 1894.

Sharpsteen, William C. "Vanished Waters of Southeastern San Francisco." *California Historical Society Quarterly* 21 (June, 1942): 113–26.

Shyrock, Henry S., Jr. "Redistribution of Population: 1940–1950." *Journal of the American Statistical Association* 46 (December, 1951): 436.

Simon, Anne W. *The Thin Edge: Coast and Man in Crisis*. New York: Harper and Row, 1978.

Singer, Joseph, ed. *History of Technology*. 10 vols. Oxford, England: Clarendon Press, 1954–58.

Skinner, Alanson. *The Indians of Greater New York*. Cedar Rapids, Iowa: Torch Press, 1915.

Smallwood, William Martin. *Natural History and the American Mind*. New York: Columbia University Press, 1941.

Smith, Frank E. *Conservation in the United States: A Documentary History*. Vol. 3, *Land and Water, Part 2: 1900–1970*. New York: Van Nostrand-Reinhold, 1971.

Smith, Norman. *Man and Water: A History of Hydro-Technology*. London: Peter Davies, 1975.

Stanton, William. *The Great United States Exploring Expedition of 1838–1842*. Berkeley and Los Angeles: University of California Press, 1975.

Starkey, Marion. *Land Where Our Fathers Died: The Settlement of the Eastern Shores of North America, 1607–1735*. London: Constable, 1964.

Stegner, Wallace. *Beyond the Hundredth Meridian*. Boston: Houghton Mifflin, 1962.

Sterling, Kier Brooks. *Last of the Naturalists: The Career of Clinton Hart Merriam*. New York: Arno Press, 1974.

Sterling, Phillip. *Sea and Earth: The Life of Rachel Carson*. New York: T. Y. Crowell, 1970.

Stilgoe, John R. "Jack-o'-Lanterns to Surveyors: The Secularization of Landscape Boundaries." *Environmental Review* 1 (January, 1976): 14–31.

Swain, Donald. *Federal Conservation Policy: 1921–1933*. Berkeley: University of California Press, 1963.

Teal, John, and Mildred Teal. *Life and Death of the Salt Marsh*. Boston: Little Brown, 1969.

—— and ——. "Ribbon of Green: The Epic of a Salt Marsh." *Audubon Magazine*, November, 1969, p. 4.

Teele, Ray Palmer. *The Economics of Land Reclamation in the United States*. New York: A. W. Shaw, 1927.

Theobald, Robert. *Habit and Habitat*. Englewood Cliffs, N.J.: Prentice-Hall, 1972.

Thoreau, Henry David. *Cape Cod*. Boston: James R. Osgood, 1871.

Trefethen, James B. *An American Crusade for Wildlife*. New York: Boone and Crockett Club, 1975.

Turner, Frederick Jackson. "The Significance of the Frontier in American History." In *The Turner Thesis Concerning the Role of the Frontier in American History*, edited by George Rogers Taylor. Lexington, Mass.: D. C. Heath, 1956.

Udall, Stewart. *The Quiet Crisis*. New York: Holt, Rinehart and Winston, 1963.

Uzes, Francois D. *Chaining the Land: A History of Surveying in California*. Ann Arbor: Edward Brothers, 1977.

Van Hise, Charles. *The Conservation of Natural Resources in the United States*. Revised edition. New York: Macmillan, 1923.

Van Veen, Johan. *Dredge, Drain and Reclaim: The Art of a Nation*. The Hague: Martinus Nijhoff, 1955.

Viereck, Phillip. *The New Land*. New York: John Day, 1967.

Wallace, Bruce, ed., *Essays in Social Biology*. 2 vols. Englewood Cliffs, N.J.: Prentice-Hall, 1972.

Walvin, James. *Beside the Seaside: A Social History of the Popular Seaside Holiday*. London: Allen Lane, 1978.

Ward, Ritchie. *Into the Ocean World: The Biology of the Sea*. New York: Alfred Knopf, 1974.

Warner, Sam Bass. *The Private City*. Philadelphia: University of Pennsylvania Press, 1968.

Weber, Eugene W. "Comprehensive River Basin Planning: Development of a Concept." *Journal of Soil and Water Conservation* (July–August, 1964): 133–38.

Welter, Rush. *The Mind of America, 1820–1860*. New York: Columbia University Press, 1975.

Weyburn, Peggy. *The Edge of Life*. San Francisco: Sierra Club Books, 1972.

White, Lynn, Jr. "The Historical Roots of Our Ecological Crisis." *Science*, March 10, 1967, pp. 1203–1207.

Whitehill, Walter Muir. *Boston: A Topographical History*. Cambridge, Mass.: Belknap Press, 1959.

Williams, Frances L. *Matthew Fontaine Maury: Scientist of the Sea*. New Brunswick: Rutgers University Press, 1963.

Woods, Barbara, ed. *Eco-Solutions: A Casebook for the Environmental Crisis*. Cambridge, Mass.: Schenkman Publishing, 1972.

Worster, Donald. *Nature's Economy: The Roots of Ecology*. San Francisco: Sierra Club Books, 1973.

————, ed. *American Environmentalism: The Formative Period, 1860–1915*. New York: John Wiley and Sons, 1973.

Wright, Thomas. "On the Mode Most Easily and Effectually Practicable of Drying Up the Marshes of the Maritime Parts of North America." *Transactions of the American Philosophical Society* 4 (1799): 243–46.

Zwick, David, and Marcy Benstock. *Water Wasteland: Nader's Study Group Report on Water Pollution*. New York: Bantam Books, 1971.

Dissertations

Hedgpeth, Joel. "Ecological and Distributional Relationships of Marine and Brackish Water Invertebrates of the Coast of Texas and Louisiana." Ph.D. dissertation, University of California, Berkeley, 1952.

Liebetrau, Suzanne Fries. "Trailblazers in Ecology: The American Ecological Consciousness, 1850–1864." Ph.D. dissertation, University of Michigan, 1973.

Van Brocklin, Ralph Merton. "The Movement for the Conservation of Natural Resources in the United States Before 1901." Ph.D. dissertation, University of Michigan, 1952.

Voss, Ellis. "Summer Resort: An Ecological Analysis of a Satellite Community." Ph.D. dissertation, University of Pennsylvania, 1941.

Government Publications and Documents

Baird, Spencer F. *Report of the United States Fish Commissioner, 1872–73*. Washington, D.C.: Government Printing Office, 1874.

Commonwealth of Massachusetts Bay. *Book of the General Laws and Libertyes of Massachusettes* (1636). Edited by Thomas Barnes. Facsimile of the 1648 edition. San Marino, Calif.: Huntington Library, 1975.

Congressional Quarterly Almanac. Washington, D.C.: Congressional Quarterly Service, various years.

Dingell, John. "Wetlands Are Not Wastelands." *Congressional Record*. Vol. 117, November 9, 1971. Washington, D.C.: Government Printing Office.

Gallatin, Albert. "Report on Roads and Canals." Miscellaneous Document No. 250, 10th Congress, 2d session, April 4, 1808. *American State Papers: Docu-*

ments, Legislative and Executive of the Congress of the United States. Vol. 7. Washington, D.C.: Gales and Seaton, 1832.

Haeckel, Ernst. "Plankton Studies" (1890). Translated by George Wilton Field. In *U.S. Fish Commissioner Report, 1889–1891*. Washington, D.C.: Government Printing Office, 1893.

Jefferson, Thomas. "Report on Fisheries and Their Decline and Foreign Competition." 1st Congress, 2d session, February 1, 1791. In *American State Papers: Documents, Legislative and Executive of the Congress of the United States*. Vol. 7. Washington, D.C.: Gales and Seaton, 1832.

Kerpleman v. *Board of Public Works of Maryland. Maryland Reporter* 261. Minneapolis, Minn.: West Publishing, 1971.

Kettleborough, Charles. *Drainage and Reclamation of Swamp and Overflowed Lands*. Indiana Bureau of Legislative Information Bulletin 2. Indianapolis: W. B. Burford, 1914.

Marsh, George Perkins. *Report Made Under the Authority of the Legislature of Vermont on the Artificial Propagation of Fish*. Burlington, Vt.: Free Press Print, 1857.

Mobius, Karl. "The Oyster and the Oyster Culture." Translated in *Report of the U.S. Fish Commission, 1880*. Washington, D.C.: Government Printing Office, 1881.

Nease, Jack, ed. *Man's Control of the Environment*. Washington, D.C.: Congressional Quarterly Service, 1970.

Shalowitz, Aaron L. *Shore and Sea Boundaries*. 2 vols. Washington, D.C.: Government Printing Office, 1962.

U.S. Congress, House. *Report on the Lands of the Arid Region of the United States*, by John Wesley Powell. Executive Document 73. 45th Congress, 2d session, April 3, 1878.

———. *Reclamation of Tidelands*, by J. O. Wright. House Document 820. 59th Congress, 2d session, December 3, 1906.

———. Committee on Merchant Marine and Fisheries. *Estuarine and Wetlands Legislation. Hearings Before the Subcommittee on Fisheries and Wildlife Conservation*. 89th Congress, 2d session, June 16, 22, 23, 1966.

———. ———. *Estuarine Areas*. House Report 989. 90th Congress, 1st session, November 28, 1967.

U.S. Congress, Senate. *The National Estuarine Pollution Study, Report of the Secretary of the Interior*. Senate Document 91–58. 91st Congress, 2d session, March 25, 1970.

———. Committee on Commerce. *Estuaries and Their Natural Resources. Hearing on H.R. 25 and S. 695*. 90th Congress, 2d session, June 4, 1968.

———. ———. *Estuarine Areas*. Senate Report 1419. 90th Congress, 2d session, July 17, 1968.

———. *Report on the Overflows of the Delta of the Mississippi*, prepared by Charles Ellet. Executive Document 2. 32d Congress, 1st session, January 21, 1852.

U.S. Council on Environmental Quality. "Man on the Seashore: An Exponential

Force against a Finite Limit." Prepared by Joel Hedgpeth. In *Wildlife and America*, edited by H. P. Brokaw. Washington, D.C.: Government Printing Office, 1978.

U.S. Department of Agriculture. "Reclamation of Salt Marshes." Farmers Bulletin No. 320. Experiment Station Work. Vol. 46. April 8, 1908. Washington, D.C.: Government Printing Office, 1908.

U.S. Department of the Interior. Fish and Wildlife Service. *Classification of Wetlands and Deepwater Habitats of the United States*. Prepared by Lewis M. Cowardin et al. Washington, D.C.: Government Printing Office, 1979.

————. ————. *National Estuary Study*. 7 vols. Washington, D.C.: Government Printing Office, 1970.

U.S. Department of the Interior. U.S. Geological Survey. "Preliminary Report on the Seacoast Swamps of the Eastern United States." Prepared by Nathaniel S. Shaler. In *U.S. Geological Survey Sixth Annual Report*. Washington, D.C.: Government Printing Office, 1886.

U.S. Fish Commission. "The First Decade of the U.S. Fish Commission." Prepared by George Browne Goode. In *Report of the U.S. Fish Commissioner, 1880*. Washington, D.C.: Government Printing Office, 1883.

————. "Proceedings and Papers of the National Fishery Congress." In *Bulletin of the U.S. Fish Commission, 1897*. Washington, D.C.: Government Printing Office, 1898.

U.S. War Department. Army Corps of Engineers. *Shore Control and Port Administration: Investigation of the Status of National, State, and Municipal Authority over Port Affairs*. Washington, D.C.: Government Printing Office, 1923.

Warren, George M. *Tidal Marshes and Their Reclamation*. U.S.D.A. Office of Experiment Stations, Bulletin 240. Washington, D.C.: Government Printing Office, 1911.

Wright, J. O. *Reclamation of Tidelands*. 59th Congress, 2d session. Doc. 820 (Experiment Stations Office), December 3, 1906.

Index